Lyotropic Liquid Crystals

Lyotropic Liquid Crystals

Ingo Dierking

Department of Physics and Astronomy,
University of Manchester

Antônio Martins Figueiredo Neto

Institute of Physics, Experimental Physics Department,
University of São Paulo

OXFORD
UNIVERSITY PRESS

OXFORD
UNIVERSITY PRESS

Great Clarendon Street, Oxford, OX2 6DP,
United Kingdom

Oxford University Press is a department of the University of Oxford.
It furthers the University's objective of excellence in research, scholarship,
and education by publishing worldwide. Oxford is a registered trade mark of
Oxford University Press in the UK and in certain other countries

Published in the United States of America by Oxford University Press
198 Madison Avenue, New York, NY 10016, United States of America

British Library Cataloguing in Publication Data
Data available

Library of Congress Control Number: 2024936405

ISBN 9780192898654

DOI: 10.1093/9780191924927.001.0001

Printed and bound by
CPI Group (UK) Ltd, Croydon, CR0 4YY

Cover image: Texture of a cholesteric lyotropic liquid crystal, observed in polarized optical microscope.
Mixture of potassium laurate/l-decanol/water/brucine sulfate, at 20°*C*. Sample in 200 μm thick microslide.
Cholesteric discotic mesophase, pitch of 125 μm. Thin black lines, parallel to the cholesteric axis, are ferrofluid linear
clusters indicating the direction of the external magnetic field (1 *kG*) applied to orient the structure.
© Antônio Martins Figueiredo Neto

Contents

1
Introduction

Liquid crystals are anisotropic fluidic systems which in a thermodynamic sense are located between the three-dimensionally ordered structure of a crystal and that of a flowing isotropic liquid. They are thus partially ordered systems which may exhibit purely orientational order of anisotropic molecules or particles, or in addition one- or two-dimensional translation order, which can be short- or long-range. Already here, we can see that there are plenty of possibilities to generate liquid crystal phases of varying features of order.

Of the two main classes of liquid crystals the thermotropic ones are those that are normally mentioned and introduced, often exclusively, due to their applicational importance in displays, flat panel screens, and electro-optic components. As the name implies, in thermotropic liquid crystals the liquid crystalline phases are observed on change of temperature; the phase with the highest order and thus lowest symmetry is located at low temperatures, while successively becoming less ordered, with higher symmetry, as the temperature is raised. Thermotropic liquid crystals are also the ones to date the discovery of liquid crystallinity to 1888, when Friedrich Reinitzer, a botanist at the University of Prague, observed the presumed occurrence of 'two melting points' for cholesteryl-benzoate on heating [1]. What he described in his publication was the transition from a crystalline substance to its liquid crystalline state, a cloudy, milky white appearance, of a flowing liquid with an iridescent colour play due to light scattering and selective reflection from a helical superstructure. At higher temperature a subsequent transition into a clear liquid was observed, the so-called second melting point. It was one year later that Otto Lehmann, a specialist in polarized optical microscopy at the University of Karlsruhe, realized that the state between the crystal and the clear liquid was in fact a completely novel state of matter, which he named 'liquid crystal' [2].

Up to the present day, a zoo of thermotropic liquid crystalline phases has been described, each varying from the others by some features of order. The simplest of these phases is the nematic phase which exhibits solely orientational order of a preferred direction called the director **n**, while the molecular centres of mass are isotropically distributed. The degree of order in its simplest form is described by a scalar order parameter S [3], where

$$S = \frac{1}{2} \langle 3 \ cos^2 \vartheta_i - 1 \rangle$$

The brackets <> denote the temporal and spatial average and ϑ_i is the angle between the long molecular axis of molecule i with the director **n.** S is temperature dependent with no order, S=0, in the isotropic phase, and S=1 for perfect orientational order.

Lyotropic Liquid Crystals. Ingo Dierking and Antônio Martins Figueiredo Neto, Oxford University Press.
 DOI: 10.1093/9780191924927.003.0001

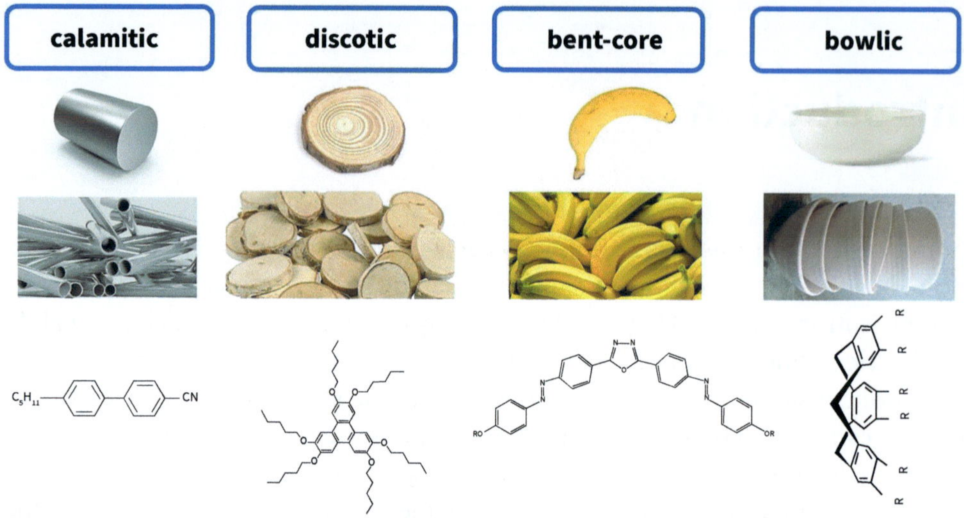

Figure 1.1 Different anisotropic molecular shapes lead to the formation of different thermotropic liquid crystal phases (mesophases). General classes that are distinguished are calamitic, discotic, bent-core, and bowlic mesogens.

Experimentally, the nematic phase shows values of about S≈0.3 at the first-order Iso-N transition and S≈0.6–0.7 at low temperatures.

The molecules themselves are often rod-shaped (calamitic), disc-shaped (discotic), of bent nature (bent-core or banana molecules), or bowl-shaped (bowlic), just to name the simplest of the molecular geometric forms, as shown in Figure 1.1. On reducing temperature, most often the fluid smectic phases are formed from the nematic (or isotropic) phase. These exhibit a one-dimensional positional order in addition to orientational order, thus the formation of smectic layers. The fluid smectic phases are designated as smectic A if the director is parallel to the smectic layer normal and smectic C if it is inclined by an angle θ, called the tilt angle. The latter is generally a function of temperature and the smectic A to smectic C transition is most often of second order. Further cooling would then result in the hexatic smectic phases where short-range hexatic positional order within a smectic layer is formed, so called bond-orientational order. These are called smectic B for the orthogonal phase, smectic I when tilted to the apex and smectic F when tilted to the side of the hexagon. The thermodynamic scheme of the simplest of the thermotropic phases is depicted in Figure 1.2. Further phases which occur on cooling will then exhibit long-range positional order and are also called soft crystal phases.

All in all, there are approximately 25–30 different thermotropic phases[4–6], including (i) those shown in Figure 1.2, (ii) the discotic phases [7], which are nematic and columnar, (iii) a variety of novel bent-core phases [8], which are

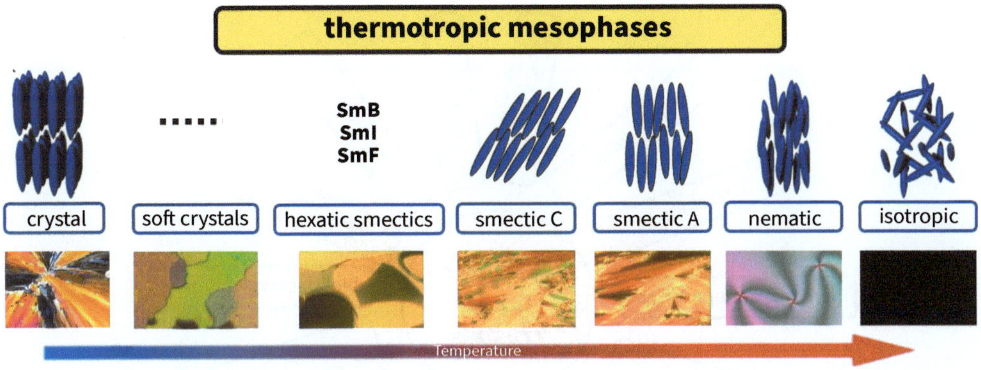

Figure 1.2 Calamitic molecules are often used to illustrate the complexity and diversity of the thermotropic phases formed. Also here, several different groups of order phenomena are distinguished: (i) pure orientational order leading to nematic phases; (ii) additional one-dimensional positional order exhibiting the fluid smectic phases A and C; (iii) in plane hexatic order, also called bond-orientational order, leading to the hexatic smectic phases; and (iv) the three-dimensionally ordered soft crystal phases.

nematic and mainly smectic, and (iv) some bowlic phases [9] of columnar nature. The addition of chirality [10], either within the mesogens or via chiral dopants, can lead to additional phases, such as the Blue Phases (BP), which are located between the isotropic and the chiral nematic phase, the Twist Grain Boundary phases (TGB), located between the chiral nematic and the fluid smectic phases, and a variety of ferroelectric, antiferroelectric, and intermediate phases [11,12]. Given the enormous importance of thermotropic liquid crystals for applications and in industry [13–16], these are the primarily studied and investigated materials and the larger amount of the available literature is dedicated to those materials [17,18].

Yet, there is a whole class of liquid crystalline systems, which is equally as diverse and is attracting increasing interest both from the viewpoint of fundamental science as well as applications. These are the lyotropic liquid crystals [19,20], phases that are formed by the addition of anisotropic molecules or materials to an isotropic solvent, often water. The dispersed components can be of very different nature, for example amphiphilic molecules that aggregate to larger (anisotropic) units [21]. Some mesogens even exhibit both thermotropic and lyotropic behaviour, a group which are called amphotropic liquid crystals [22]. But also inorganic dispersants [23] like the needle-shaped nanoparticles of vanadium pentoxide [24], V_2O_5, the plate-like clays [25, 26] or graphene oxide [27–31] are seen to exhibit liquid crystal properties. Furthermore, synthesized nanostructures such as nanorods [32–34] or modified nanotubes [35,36] show lyotropic phases, as do biological structures such as tobacco mosaic viruses [37–39], DNA [40], and RNA [41] or cellulose[42],

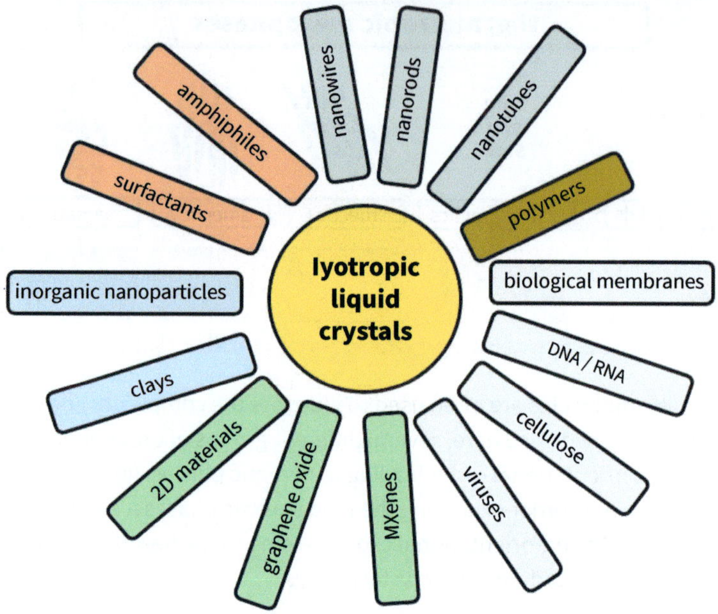

Figure 1.3 Lyotropic phases are always at least two-component materials, formed by the addition of molecules or anisotropic materials to an isotropic solvent. Common lyotropic liquid crystals are those of amphiphiles or surfactants in water, but many other scenarios are known, such as inorganic liquid crystals, clays, 2D materials in solvents, biological systems in water, elongated nanomaterials in isotropic liquids, or polymers in a solvent, just to name a selection of possible systems.

and their nanocrystals[43]. The possibilities to generate lyotropic liquid crystals are thus plentiful [44] as schematically shown in Figure 1.3, but they do have one feature in common, which separates them from thermotropics: besides temperature and pressure, concentration has become an additional variable of state (Figure 1.4).

Like in the world of thermotropic liquid crystals, chirality, i.e., the lack of mirror symmetry, adds interesting aspects also to lyotropics. These are rooted in the formation of frustrated phases, helical superstructures, and the observation of chirality-related effects, like ferroelectricity. Here, symmetry breaking can be achieved via various ways. At first, molecular chirality can be introduced, for example, through asymmetrically substituted carbon atoms in lyotropic molecules (Figure 1.5(a)). Similarly, homochiral constituents can be used in the formation of the lyotropic phase, like, for example, the cellulose nanocrystals (CNC) which are used in the production of photonic films from helical chiral nematic (cholesteric) lyotropic phases (Figure 1.5(b)). At last, one can also add chirality to the isotropic solvent, a popular example being the two chiral configurations of mandelic acid (Figure 1.5(c)).

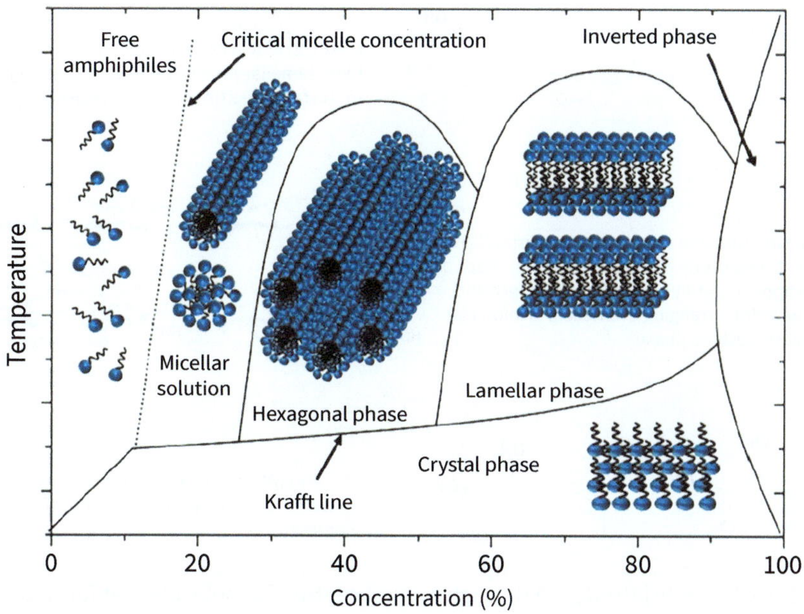

Figure 1.4 Typical schematic phase diagram of a lyotropic liquid crystal formed by amphiphilic molecules in an isotropic solvent, as a function of amphiphile concentration at constant pressure (reproduced from [44]). Different phases are mainly formed as a function of concentration, while a variation of temperature only plays an insignificant role. At the critical micelle concentration the aggregation of free amphiphiles into micelles is observed which then further aggregate into different liquid crystal phases, first in abundance of solvent and later as inverted phases at high amphiphile concentration. Often, also cubic phases appear, which can be observed at different positions in the phase diagram.

Chirality defines itself through the absence of mirror symmetry and is an inherent property of many natural systems from particle physics to biomolecules and macroscopic objects, from catalysis to stereochemistry and drug delivery [45–47]. In liquid crystals, chirality can give rise to the occurrence of novel phases and structures, such as frustrated phases and helical superstructures. In addition, these can lead to novel properties that are not observed for the achiral material, such as photonic optical effects, the selective reflection of circular polarized light of a particular wavelength, or polar structures exhibiting piezo- or ferroelectricity. The handedness of any helical structures or the sign of any observed polarization are not directly correlated to the molecular configuration. Yet, chiral properties reverse their sign when reversing image and mirror image, while non-chirality related properties, such as melting point, viscosity, or birefringence are not influenced by a change of molecular configuration.

The lyotropic liquid crystal phases and their properties outlined in this introductory chapter are the topic of the present book. In particular, we will focus on the more novel aspects which are roughly presented in Figure 1.3.

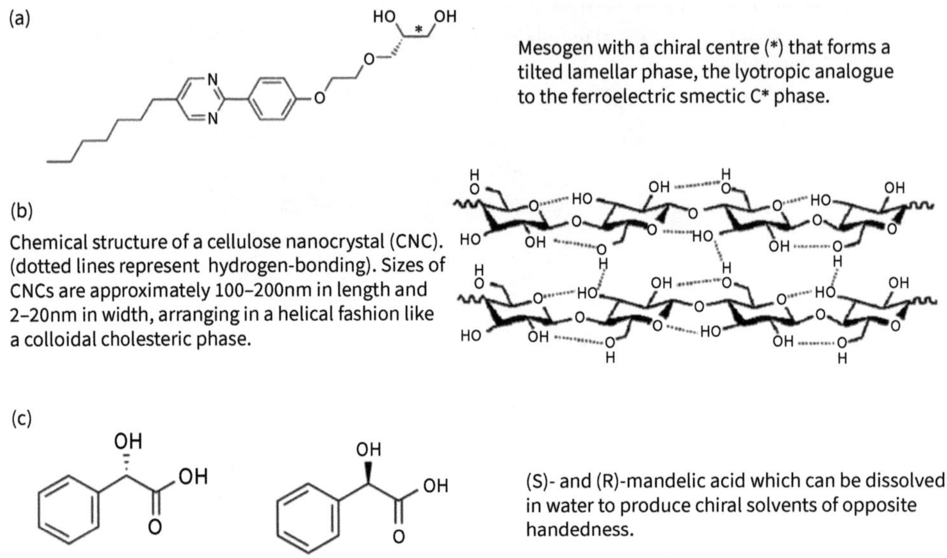

Figure 1.5 (a) Chirality introduced through a chiral centre of a molecule that forms a titled lamellar phase, in analogy to a ferroelectric chiral smectic C* phase. (b) Chirality introduced through homochiral anisotropic colloidal particles like cellulose nanocrystals (CNCs). The CNCs form a helical superstructure, leading to photonic effects similar to that of cholesterics. (c) Soluble chiral molecules can be used to generate chirality via the chiral isotropic solvent of the lyotropic liquid crystal.

References

[1] F. Reinitzer (1888). Beiträge zur Kenntniss des Cholesterins. *Monatsh. Chemie*, 9: 421.

[2] O. Lehmann (1889). Über fliessende Krystalle. *Z. Phys. Chem.*, 4: 462.

[3] V. Tsvetkov (1942). Über die Molekülanordnung in der anisotropy-flüssigen Phase. *Acta Physicochim. (USSR)*, 16: 132.

[4] P. G. de Gennes and J. Prost (1993). *The Physics of Liquid Crystals.* 2nd ed. Oxford: Clarendon Press.

[5] S. Chandrasekhar (1992). *Liquid Crystals.* Cambridge: Cambridge University Press.

[6] I. Dierking (2003). *Textures of Liquid Crystals.* Weinheim: Wiley-VCH.

[7] S. Kumar (2011). *Chemistry of Discotic Liquid Crystals: From Monomers to Polymers.* Boca Raton: CRC Press.

[8] H. Takezoe and A. Eremin (2017). *Bent-shaped Liquid Crystals: Structures and Physical Properties.* Boca Raton: CRC Press.

[9] L. Wang, D, Huang, L. Lam, and Z. Cheng (2017). Bowlics: history, advances, applications. *Liq. Cryst. Today*, 26: 85.

[10] H.-S. Kitzerow and Ch. Bahr (eds) (2001). *Chirality in Liquid Crystals.* Berlin: Springer.

[11] S. T. Lagerwall (1999). *Ferroelectric and Antiferroelectric Liquid Crystals.* Weinheim: Wiley-VCH.

[12] I. Musevic, R. Blinc, and B. Zecks (2000). *The Physics of Ferroelectric and Antiferroelectric Liquid Crystals.* Singapore: World Scientific.

[13] E. Lueder (2010). *Liquid Crystal Displays: Addressing Schemes and Electro-Optical Effects.* 2nd ed. Chichester: Wiley.

[14] S.-T. Wu and D.-K. Yang (2001). *Reflective Liquid Crystals Displays.* Chichester: Wiley; 2001.

[15] A. Y.-G. Fuh (2015). *Liquid crystals for non-display applications.* In: Lee, C. C. (ed.) The Current Trends of Optics and Photonics. Topics in Applied Physics, vol 129. Dordrecht: Springer, pp 321–336.

[16] I. Abdulhalim (2011). Non-display bio-optic applications of liquid crystals. *Liq. Cryst. Today*, 20: 44.

[17] P. J. Collings and J. W. Goodby (2020). *Introduction to Liquid Crystals.* Boca Raton: CRC Press.

[18] J. W. Goodby, P. J. Collings, T. Kato, C. Tschierske, H. Gleeson, and P. Raynes (eds) (2014). *Handbook of Liquid Crystals*: 8 Volume Set, 2nd ed. Weinheim: Wiley-VCH.

[19] A. M. F. Neto and S. R. A. Salinas (2005). *The Physics of Lyotropic Liquid Crystals: Phase Transitions and Structural Properties.* Oxford: Oxford University Press.

[20] A. G. Petrov (1999). *The Lyotropic State of Matter: Molecular Physics and Living Matter Physics.* Amsterdam: CRC Press.

[21] C. Fong, T. Le, and C. J. Drummond (2012). Lyotropic liquid crystal engineering—ordered nanostructured small molecule amphiphile self-assembly materials by design. *Chem. Soc. Rev.* 41: 1297.

[22] C. Tschierske (2002). Amphotropic liquid crystals. *Current Opinion in Colloid & Interface Science*, 7: 355.

[23] A. S. Sonin (1998). Inorganic lyotropic liquid crystals. *J. Mater. Chem.*, 8: 2557.

[24] H. Diesselhorst and H. Freundlich (1915). Über die Doppelbrechung des Vanadinpentoxydsols. *Phys. Z.*, 16: 419.

[25] E. Paineau, K. Antonova, C. Baravian, I. Bihannic, P. Davidson, I. Dozov, M. Impéror-Clerc, P. Levitz, A. Madsen, F. Meneau, et al. (2009). Liquid-crystalline nematic phase in aqueous suspensions of a disk-shaped natural beidellite clay. *J. Phys. Chem. B*, 113: 15858.

[26] A. S. Sonin, N. A. Churochkina, A. V. Kaznacheev, and A. V. Golovanov (2018). Liquid crystals of clay dispersions. *Colloid Journal*, 80: 629.

[27] Z. Xu and C. Gao (2011). Graphene chiral liquid crystals and macroscopic assembled fibres. *Nature Communications*, 2: 571.

[28] J. E. Kim, T. H. Han, S. H. Lee, J. Y. Kim, C. W. Ahn, J. M. Yun, S. O. Kim, and J. Lee (2011). Graphene oxide liquid crystals. *Angew. Chem. Int. Ed.*, 50: 3043.

[29] R. Narayan, J. E. Kim, J. Y. Kim, K.-E. Lee, and S. O. Kim (2016). Graphene oxide liquid crystals: discovery, evolution and applications. *Adv. Mater.*, 28: 3045.

[30] S. P. Sasikala, J. Lim, I. H. Kim, H. J. Jung, T. Yun, T. H. Han, and S. O. Kim (2018). Graphene oxide liquid crystals: a frontier 2D soft material for graphene-based functional materials. *Chem. Soc. Rev.*, 47: 6013.

[31] A. Draude and I. Dierking (2019). Lyotropic liquid crystals from colloidal suspensions of graphene oxide. *Crystals*, 9: 455.

[32] S. Zhang, P. W. Majewski, G. Keskar, L. D. Pfefferle, and C. O. Osuji (2011). Lyotropic self-assembly of high-aspect-ratio semiconductor nanowires of single-crystal ZnO. *Langmuir*, 27: 1616.

[33] Z. Ren, C. Chen, R. Hu, K. Mai, G. Qian, and Z. Wang (2012). Two-step self-assembly and lyotropic liquid crystal behavior of TiO_2 nanorods. *J. Nanomater.*, 2012; 180989. https://doi.org/10.1155/2012/180989.

[34] L.-S. Li, J. Walda, L. Manna, and A. P. Alivisatos (2002). Semiconductor nanorod liquid crystals. *Nano Lett.*, 2: 557.

[35] W. Song, I. A. Kinloch, and A. H. Windle (2003). Nematic liquid crystallinity of multiwall carbon nanotubes. *Science*, 302: 1363.

[36] S. Badaire, C. Zakri, M. Maugey, A, Derré, J. N. Barisci, G. G. Wallace, and P. Poulin (2005). Liquid crystals of DNA-stabilized carbon nanotubes. *Adv. Mater.*, 17: 1673.

[37] F. C. Bawden, N. W. Pirie, J. D. Bernal, and I. Fankuchen (1936). Liquid crystalline substances from virus-infected plants. *Nature*, 138: 1051.

[38] S. Fraden, G. Maret, and D. L. D. Caspar (1993). Angular correlations and the isotropic-nematic phase transition in suspensions of tobacco mosaic virus. *Phys. Rev. E*, 48: 2816.

[39] Z. Dogic and S. Fraden (2006). Ordered phases of filamentous viruses. *Current Opinion in Colloid & Interface Science*, 11: 47.

[40] A. Leforestier and F. Livolant (2010). The bacteriophage genome undergoes a succession of intracapsid phase transitions upon DNA ejection. *J. Mol. Biol.*, 396: 384.

[41] G. Zanchetta, M. Nakata, M. Buscaglia, N. A. Clark, and T. Bellini (2008). Liquid crystal ordering of DNA and RNA oligomers with partially overlapping sequences. *J. Phys. Condens. Matter*, 20: 494214.

[42] P. Zugenmaier (2008). Crystalline Cellulose and Derivatives: Characterization and Structures. Berlin: Springer.

[43] D. G. Gray and X. Mu (2015). Chiral nematic structure of cellulose nanocrystal suspensions and films; polarized light and atomic force microscopy. *Materials*, 8: 7873.

[44] I. Dierking and A. M. F. Neto (2020). Novel trends in lyotropic liquid crystals. *Crystals*, 10: 604.

[45] R. Janoschek (ed.) (1991). Chirality: From Weak Bosons to the α-helix. Berlin: Springer.

[46] G. H. Wagniere (2007). On Chirality and the Universal Asymmetry. Zürich: Wiley-VCH.

[47] D. B. Amabilino (ed.) (2009). Chirality at the Nanoscale. Weinheim: Wiley-VCH.

2
Theory

2.1 Introduction

In this chapter we will revise some basic aspects of the Onsager and the Landau–de Gennes theories to describe the nematic to isotropic and nematic to nematic phase transitions. These theories are mean-field theories and catch the main physical aspects of these transitions. It is not our aim here to repeat the mathematical calculations of these theories, since there is a good bibliography available that covers this. In particular, good books can be consulted by readers with detailed descriptions of these theories and extensive bibliography [1–4].

2.2 Onsager theory for plates and rods

Let us assume a gas of particles of nanometric dimensions, with the shape of rods. This system has two entropies, one associated with the translation degrees of freedom of the particles' centre of mass and another associated with the rotations around axes passing through the centre of mass. The stability of the system depends on the coupling between these entropies that is directly related to the concentration of particles. At this point we need the concept of *excluded volume*. As a working example, consider two cylindrical particles represented by hard rods, one of them being fixed in the reference frame axes. The excluded volume is defined as the volume around the fixed rod where the centre of mass of the second rod cannot be located, due to the impenetrability of the particles. It is straightforward to see that the excluded volume is bigger when the axes of the rods are perpendicular to each other, and is smaller when they are parallel. The parallel orientation of the rods favours the translational entropy, but the orientational entropy has a tendency to minimize. The stable organization of the rods, at a given concentration of particles, is achieved in the process of maximization of both entropies. In the limit of a very small numerical density of particles (ρ) the isotropic phase of non-oriented rods is the more stable, maximizing both entropies. On the other side, in the case of closed-packed rods, the parallel orientation of the particles stabilizes a phase with the axes of the rods parallel to each other. So, a transition between these two phases is expected as a function of the numerical density of particles.

Onsager proposed in 1949 an elegant theory taking into account the arguments briefly described above [5] to account for the nematic to isotropic phase transition in a system composed by rod-like and plate-shaped particles. He observed the similarity between an ideal gas and an ideal solution of particles. Rods have a length

Lyotropic Liquid Crystals. Ingo Dierking and Antônio Martins Figueiredo Neto, Oxford University Press.
 DOI: 10.1093/9780191924927.003.0002

(L) much larger than their diameter (D). In the case of platelets, the dimension in the platelet plane is much larger than their thickness. In his model, the classical statistical mechanics approach is employed to calculate the partition function of a system of anisotropic particles, subjected to a pairwise steric repulsive potential between particles i and j, v_{ij}. This potential depends on the distance between particles and their relative orientation. It is possible to consider electrically charged particles in an ionic medium, keeping the neutrality of the solution. The counterions shield the particles' charges and, according to Onsager's proposal, it is enough to consider rods with a larger diameter. As $V_1 = (\pi D^2 L)/4$ is the volume of a rod, it is assumed that the volume fraction of particles (ϕ) is $\phi = \rho\ V_1 << 1$.

With the partition function calculated in the mean-field framework, the free energy of the system can be obtained and minimized by using the Euler–Lagrange equation [2]. An equivalent (and elegant) way to obtain the free energy of the system is to consider initially the case of a dilute gas of hard spheres (where only the translational degrees of freedom exist), and then introducing the additional degrees of freedom related to the orientational-distribution function (f_a) of the rods [1]. The numerical density of rods pointing in the solid angle $d\Omega$ around the direction defined by the unit vector $\hat{a}$, is $\rho f_a d\Omega$, with $\int f_a\, d\Omega = 1$. Choosing z, from the three orthogonal reference frame axes, as the preferred direction and θ the angle of the rod's axis with respect to z, the uniaxial order parameter S is defined as [2]:

$$S \equiv \int P_2\,(cos\theta)\, f_a\,(cos\theta)\ d\,(cos\theta), \tag{2.1}$$

where $P_2(cos\theta) = (3cos^2\theta - 1)/2$ is the second-order Legendre polynomial. The maximization of the entropy or, equivalently, minimization of the free energy leads to a first-order transition between the uniaxial nematic to the isotropic phase at a critical concentration ρ_c . For symmetry reasons, considering the local and global symmetry of the nematic and isotropic phases, a first-order transition is expected, with a jump in the order parameter at ρ_c [3]. Figure 2.1 sketches the order parameter as a function of ρ. For $\rho < \rho_c$ the isotropic phase is stabilized and, on the other hand, for $\rho > \rho_c$ the uniaxial nematic phase is stabilized. At $\rho = \rho_c$ there is a coexistence of both phases, as expected in a first-order phase transition. The jump in the order parameter predicted by this theory is $\Delta S \sim 0.7$, differently from what is typically observed in experiments with thermotropic liquid crystals ($\Delta S \sim 0.4$). Considering the assumptions made to derive the total entropy and the free energy of the system, Onsager's theory is not bad. Certainly, short-range interactions between particles, different from the excluded-volume one, should be considered to account for actual systems.

Onsager's ideas inspired researchers to explore new particle shapes as, e.g., the biaxial one. Tjipto-Margo and Evans [6] described the isotropic to nematic phase transition in a dense fluid composed of hard biaxial ellipsoids. The interesting result obtained is that the first-order nature of the transition is reduced, when compared

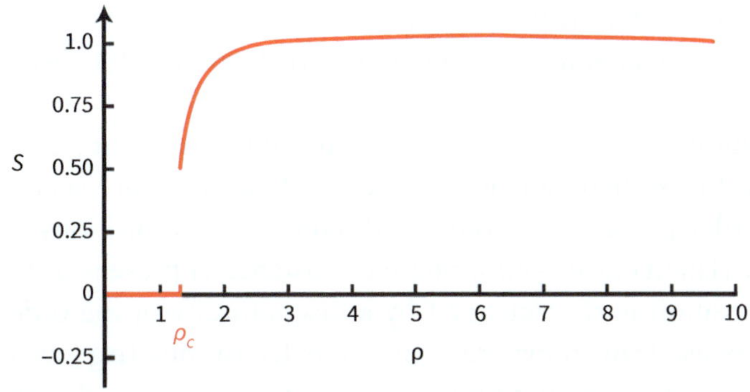

Figure 2.1 Sketch of the result of the maximization of the total entropy of a system of rod-shaped particles, showing the orientational order parameter as a function of the numerical density of particles.

to the case of uniaxial particles. In the case of thermotropics presenting the nematic to isotropic phase transition, many actual nematogens have shapes that cannot be simply approximated by a plate or a rod. So, this extension of Onsager's theory could be applicable. This aspect is particularly important in lyotropics since micelles, in some conditions, show orthorhombic symmetry. We will come back to this point in later sections.

Up until this point a collection of identical particles has been considered. However, the theory can be adapted to a set of polydisperse particles. Assuming rod-like particles with fixed diameter and a log-normal length distribution with a finite upper cutoff, Speranza and Sollich [7] investigated the isotropic–nematic phase coexistence and, interestingly, an isotropic–nematic–nematic coexistence region. From the experimental point of view, this three-phase coexistence had been observed in a dispersion of organophilic boehmite rods [8].

More recently, Xiao and Sheng generalized Onsager's theory to take into account rod particles of small aspect ratio (defined as L/D) [9]. Besides the second virial coefficient present in Onsager's Helmholtz free energy, high-order virial coefficients were considered. Molecular dynamic simulations have shown the importance of considering the short-range order of particles besides the steric repulsion between them for particles with small aspect ratio.

2.3 The Landau theory of phase transitions

In 1937 Landau published a seminal paper [10] that inspired and continues to inspire physicists working in condensed matter Physics around the world. The Landau theory is phenomenological. It does not need to consider the specific forms of the interatomic and intermolecular microscopic interactions that are, in essence, ultimately responsible for the phase transitions, but only their symmetries. The theory describes transitions where there is a symmetry change. The concepts of symmetry and symmetry breaking are the basis of the Landau theory, which allows its application in several branches of Physics, and attribute to the theory a universal character.

Phase transitions that imply a change in the symmetry and, consequently, in the entropy of the system can be classified as first-order or second-order transitions. First-order phase transitions are discontinuous transitions, accompanied by a release of (latent) heat, and a phase coexistence is present in the phase diagram. This transition is characterized by a discontinuity of the order parameter. Second-order phase transitions are continuous transitions (e.g., as a function of temperature) and an anomaly is present in the specific heat. In this case, the order parameter vanishes continuously when passing from the less to the more symmetric phase.

Two concepts are particularly important in the theory: the free-energy density ($f = F/V$) and the order parameter, where F is essentially the Helmholtz or the Gibbs free energy and V the volume of the system. f may depend on different parameters, e.g., temperature, pressure, chemical composition, external fields, and on the order parameters relevant for the phase transition. Landau postulated that, in the vicinity of a second-order phase transition, f can be written as an expansion (power series) in terms of the invariants of the order parameter and its spatial derivatives. The coefficients of this expansion are functions of temperature. The critical order parameter (non-critical order parameters will be discussed later) of a particular system that characterizes the phase transition may be a scalar, a vector, or even a tensor (as in the case of liquid crystals). So, the obtention of the invariants of the order parameter is necessary to properly write down the free energy of the system. The Landau theory is, essentially, a mean-field theory and was shown to be easily generalized to account for first-order phase transitions.

Let us assume a system that presents a second-order phase transition at $T = T_c$, characterized by the invariant of the order parameter represented by ϵ. For $T > T_c$ we have $\epsilon = 0$ and for $T < T_c$, $\epsilon \neq 0$. The free-energy density is written [2] in the vicinity of T_c as Eq. 2.2 that keeps just even powers of ϵ due to the symmetry of the physical system (i.e., the free-energy density must be invariant under the $\epsilon \rightarrow -\epsilon$ transformation):

$$f(\epsilon, T) \simeq f_0[T] + \frac{1}{2}A(T)\epsilon^2 + \frac{1}{4}C(T)\epsilon^4 + \frac{1}{2}D[\vec{\nabla}\epsilon(\vec{r})]^2 .... \tag{2.2}$$

where $f_0[T] \equiv f[\epsilon_0(T), T]$ is an analytic function of T, with $\epsilon_0(T)$ being the value of ϵ that minimizes $f(\epsilon, T)$ at the temperature T; $A = a(T - T_c)$, a, C, and D are positive constants, known as Landau coefficients; and the last term in Eq. 2.2 takes into account the spatial variations of ϵ.

To account for first-order phase transitions, coefficient C assumes negative values and additional orders in the expansion may be needed (e.g., $\frac{1}{6}E(T)\epsilon^6$) to keep the positive concavity of the free-energy density. If the symmetry of the system allows, odd terms are included in the expansion of f, and Eq. 2.2 assumes the form [2]:

$$f(\epsilon, T) \simeq f_0[T] + \frac{1}{2}A(T)\epsilon^2 + \frac{1}{3}B(T)\epsilon^3 + \frac{1}{4}C(T)\epsilon^4 + \frac{1}{2}D[\vec{\nabla}\epsilon(\vec{r})]^2.... \tag{2.3}$$

2.4 The Landau–de Gennes theory for the isotropic–nematic and nematic–nematic transitions

The uniaxial nematic phase is characterized by local director $\hat{n}(\vec{r})$, defined as the mean orientation of the basic unities of the system (e.g., cylindrical molecules), located in a small volume at position $\vec{r}$. Given θ is the angle between the long axis of the molecules and the director, the scalar order parameter is defined as [2]:

$$S(\vec{r}) = \ll P_2(cos\theta) \gg, \tag{2.4}$$

where $\ll\ \gg$ stands for a spatial averaging of the molecular orientations at $\vec{r}$ and is the second Legendre polynomial. In the uniaxial–nematic phase the states corresponding to $\hat{n}(\vec{r})$ and $-\hat{n}(\vec{r})$ are physically equivalent.

The convenient order parameter to describe the nematic to nematic and the nematic to isotropic phase transitions is a second-rank symmetric traceless tensor $Q_{\alpha\beta}$ [1]. In principle, any static response function of the liquid crystalline system which shows its symmetry (e.g., magnetic susceptibility, dielectric constant) may be used to write **Q**. In uniaxial nematics it is usual to orient the director parallel to the z-axis of the laboratory frame axes (1, 2, 3 equivalent to x, y, z). Since we need the order parameter to vanish at the isotropic phase, it is necessary to remove the anisotropic part of the tensor and, in terms of the dielectric tensor, we have:

$$Q_{ij} = S(\vec{r})[3n_i(\vec{r})n_j(\vec{r}) - \delta_{ij}], \tag{2.5}$$

and

$$\mathbf{Q}(\vec{r}) = S(\vec{r})\begin{pmatrix} -1 & 0 & 0 \\ 0 & -1 & 0 \\ 0 & 0 & 2 \end{pmatrix}. \tag{2.6}$$

Note that the trace of the tensor is zero. The free-energy density is written as [2]:

$$f \simeq f_0[T] + \frac{1}{2}a(T - T_c^*)Q_{ij}(\vec{r})Q_{ij}(\vec{r}) + \frac{1}{3}BQ_{ij}(\vec{r})Q_{j_k}(\vec{r})Q_{ki}(\vec{r})$$
$$+ \frac{1}{4}C_1[Q_{ij}(\vec{r})Q_{j_i}(\vec{r})]^2 + \frac{1}{4}C_2Q_{ij}(\vec{r})Q_{j_k}(\vec{r})Q_{kl}(\vec{r})Q_{li}(\vec{r}) + \frac{1}{2}L_1\partial_iQ_{j_k}(\vec{r})\partial_iQ_{j_k}(\vec{r})$$
$$+ \frac{1}{2}L_2\partial_iQ_{ij}(\vec{r})\,\partial_kQ_{kj}(\vec{r}).... \tag{2.7}$$

where T_c^* is a temperature just below T_c and B, C_1, C_2, L_1, L_2 are constants (repeated indices indicate summation).

An elegant (and equivalent) way to write the free-energy density, disregarding the gradient terms, was proposed by Galerne and Macerou [11] in terms of the invariants of the dielectric-tensor order parameter ($\boldsymbol{\varepsilon}(\vec{r})$), σ_1, σ_2, and σ_3, defined as:

$$\begin{aligned} \sigma_1 &= \varepsilon_{11} + \varepsilon_{22} + \sigma_{33} = 0; \\ \sigma_2 &= \frac{2}{3}(\varepsilon_{11}^2 + \varepsilon_{22}^2 + \varepsilon_{33}^2); \\ \sigma_3 &= 4\varepsilon_{11}\varepsilon_{22}\varepsilon_{33}. \end{aligned} \tag{2.8}$$

The free-energy density is:

$$f(T) = a\sigma_2 + b\sigma_3 + \frac{1}{2}c\sigma_2^2 + d\sigma_2\sigma_3 + \frac{1}{2}e\sigma_3^2, \tag{2.9}$$

where a and b are proportional to the temperature T; c, d, and e are constants. This parametrization of the free-energy density is particularly useful since the diagonal elements of $\varepsilon(\vec{r})$ can be written in terms of the birefringences of the mesophase and are experimentally accessible. These elements are written as [11]:

$$\begin{aligned} \varepsilon_{11} &= -\frac{4}{3}\langle n\rangle\left((n_2 - n_1) + \frac{n_3 - n_2}{2}\right); \\ \varepsilon_{22} &= \frac{2}{3}\langle n\rangle\left((n_2 - n_1) - (n_3 - n_2)\right); \\ \varepsilon_{33} &= \frac{4}{3}\langle n\rangle\left(\frac{(n_2 - n_1)}{2} + (n_3 - n_2)\right), \end{aligned} \tag{2.10}$$

where $n_i, i = 1, 2, 3$ are the indices of refraction along the three orthogonal axes of the laboratory frame and $\langle n\rangle$ is the mean index of refraction. The birefringences are written as $\Delta n = (n_2 - n_1)$ and $\delta n = (n_3 - n_2)$.

Equation 2.7 (without the gradient terms) and Eq. 2.9 describe the nematic to isotropic phase transition and also the biaxial to uniaxial phase transition. Two uniaxial nematic phases are predicted and one biaxial nematic phase. In particular, the formalism based on the invariants of the dielectric-tensor order parameter furnishes that in the uniaxial phase $\sigma_3 = \pm\sigma_2^{3/2}$ and in the biaxial phase $-\sigma_2^{3/2} < \sigma_3 < \sigma_2^{3/2}$ [11]. For symmetry reasons, the transition between the isotropic to the nematic phase is of first order and those between the nematic uniaxial to the nematic biaxial is of second order. The typical phase diagram in terms of the temperature T and a convenient Landau parameter c (we will come to this point later), in the vicinity of the Landau point, is depicted in Figure 2.2. This point represents the convergence of two second-order transition lines in a first-order transition line.

As stated before, this theory describes conveniently the nematic to isotropic and nematic to nematic phase transitions in the vicinity of the transition temperature.

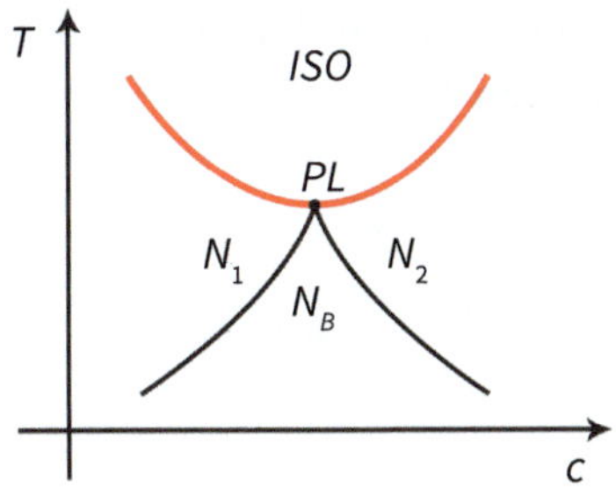

Figure 2.2 Typical partial phase diagram predicted by the Landau–de Gennes theory in terms of the temperature (T) and a convenient Landau coefficient (c). ISO, N_1, N_2, and N_B represent the isotropic phase, a uniaxial nematic phase, another uniaxial nematic phase, and the biaxial nematic phase, respectively. PL is the Landau point.

However, when experimental partial phase diagrams of lyotropic mixtures showing the three nematic phases [12,13] were explored, the behaviour predicted in Figure 2.2, with the increase of the biaxial nematic phase domain as we move away from the Landau point, is not fully observed. Moreover, an isotropic phase is observed at lower temperatures, besides the one at higher temperatures [12]. At this point, additional aspects were necessary to be considered in the theory to account for the experimental observations. The next section will be devoted to this problem.

2.5 Topological metamorphosis in the standard phase diagram

A striking characteristic of the micelles in lyotropic mixtures presenting the nematic phases is that their shape anisotropy changes as a function of temperature. These modifications were experimentally observed [14]. In thermotropic liquid crystals this aspect is expected to be negligible in most of the cases and in the temperature range where the nematic phase is observed. On the other hand, the aggregation number (number of molecules in a micelle) and the packing may be influenced by the mixture temperature.

These modifications of the shape anisotropy of the basic unit of the lyotropic nematic mesophase as a function of the temperature (and even of the type and relative concentrations of the mixture compounds) can be introduced in the theory by considering an additional non-critical order parameter τ, that will account for this aspect of the problem. It expresses the change in the micellar shape anisotropy and produces a topological metamorphosis of the standard theoretical phase diagram (Figure 2.2). τ reconstructs the standard phase diagram, imposing a symmetric folding of the existing phases and singularities [15].

To follow the development of the theory described in [15] we will parametrize the tensor order parameter and its invariants, considering r and θ as the polar coordinates in the order-parameter space. The non-null order-parameter invariants in this parametrization are written as:

$$I_1 = r^2$$
$$I_2 = r^3 cos(3\theta)\,, \tag{2.11}$$

and the free-energy density:

$$f = a_1 I_1 + a_2 I_1^2 + b_1 I_2 + b_2 I_2^2 + c_{12} I_1 I_2 + ... \tag{2.12}$$

that corresponds to Eq. 2.9. At this point we introduce the noncritical order parameter τ in Eq. 2.12 and its coupling with I_1, keeping the cubic term unchanged:

$$I'_1 = r^2 + \tau^2$$
$$I'_2 = I_2 = r^3 cos(3\theta) \tag{2.13}$$
$$I'_3 = \tau.$$

In this framework, the free-energy density assumes the form:

$$f = a_1 I'_1 + a_2 I'^2_1 + b_1 I'_2 + b_2 I'^2_2 + c_1 I'_3 + c_2 I'^2_3, \tag{2.14}$$

without invariants coupling terms. The existence of I'_3 in the free-energy density induces a symmetric folding of the standard phase diagram shown in Figure 2.2, maintaining the present singularity and introducing a new one. Moreover, the folding predicts an isotropic phase at lower temperatures, in accordance with the experimental results. Figure 2.3 shows a sketch of the phase diagram predicted by Eq. 2.14 in a convenient parametrization. The temperature T and concentration x (i.e., of the amphiphilic molecule) are linear functions of the coefficients a_1, b_1, and c_1, and were chosen to construct the phase diagram of Figure 2.3.

From the experimental point of view, lyotropic mixtures presenting the three nematic phases are composed by, at least, two amphiphilic molecules. Mixtures with just one amphiphile did not show the three nematic phases, just the uniaxial nematic. This aspect could be due to a nanosegregation of the different molecules that constitute the micelle [4]. This segregation was experimentally observed with neutron-scattering experiments in a lyotropic mixture composed of potassium laurate/decanol/water [16].

The 'take-home message' may be synthesized as: in lyotropic–nematic liquid crystals, the basic units (i.e., the micelles) change their shape anisotropy as a function of temperature and this effect must be introduced in the theory to account for the richness of the phase-diagram topology.

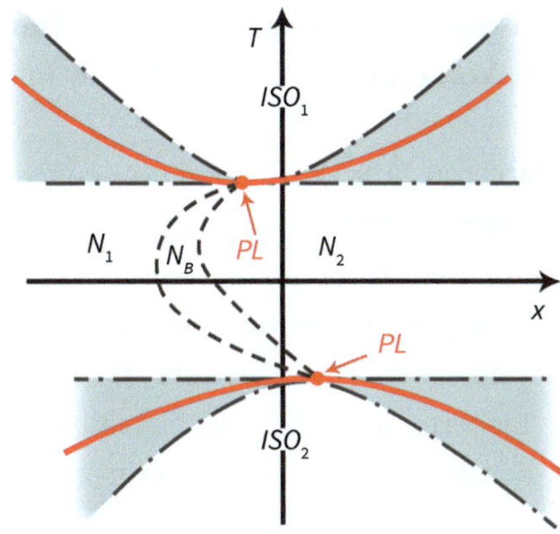

Figure 2.3 Sketch of the partial phase diagram predicted by the topological metamorphosis of the standard phase diagram, induced by the introduction of a non-critical order parameter in the free-energy density expression. ISO, N_1, N_2, N_B, and PL represent the isotropic, one uniaxial nematic, another uniaxial nematic, the biaxial phases, and the Landau point, respectively. T represents the temperature and x a convenient concentration of one of the amphiphiles of the mixture.

References

[1] P. G. de Gennes and J. Prost (1993). *The Physics of Liquid Crystals*, Oxford: Clarendon Press.

[2] P. Sheng and E. B. Priestley (1979). The Landau-de Gennes Theory of Liquid Crystal Phase Transitions, In: Priestley, EB, Wojtowicz, PJ, and Sheng, P (eds), *Introduction to Liquid Crystals*, New York: Plenum Press, pp 143–202.

[3] J. C. Tolédano and P. Tolédano (1987). *The Landau Theory of Phase Transitions.* Singapore: World Scientific.

[4] A. M. Figueiredo Neto and S. R. A. Salinas (2005). *The Physics of Lyotropic Liquid Crystals, Phase Transitions and Structural Properties.* Oxford: Oxford University Press.

[5] L. Onsager (1949). The effects of shape on the interaction of colloidal particles. *Ann. N. Y. Acad. Sci.*, 51: 627–659.

[6] B. Broto Tjipto-Margo and G. T. Evans (1991). The Onsager theory of the isotropic–nematic liquid–crystal transition: Biaxial particles in uniaxial phases. *J. Chem. Phys.*, 94: 4546–4556.

[7] A. Speranza and P. Sollich (2003). Isotropic-nematic phase equilibria in the Onsager theory of hard rods with length polydispersity. *Phys. Rev. E*, 67: 061702.

[8] P. A. Buining and H. N. W. Lekkerkerker (1993). Isotropic- nematic phase separation of a dispersion of organophilic boehmite rods. *J. Phys. Chem.*, 97: 11510–11516.

[9] X. Xiao and P. Sheng (2013). Generalized Onsager theory of liquid crystals. *Phys. Rev. E*, 88: 062501.

[10] L. Landau (1937). Translated and reprinted from Landau, L. D. 'Collected Papers' (Nauka, Moscow, 1969), Vol. 1, pp 234–252. Originally published in Zh. Eksp. Teor. Fiz. 7, pp 19–32.

[11] Y. Galerne and J. P. Marcerou (1985). Temperature-concentration behaviour of the order parameter in the nematic phases of a lyotropic liquid crystal. *J. Physique*, 46: 589–594.

[12] L. J. Yu and A. Saupe (1980). Observation of a biaxial nematic phase in potassium laurate-1-decanol-water mixtures. *Phys. Rev. Lett.*, 45: 1000–1003.

[13] E. A. Oliveira, L. Líebert, and A. M. Figueiredo Neto (1989). A new soap/detergent/water lyotropic liquid crystal with a biaxial nematic phase. *Liq. Cryst.*, 5: 1669–1675.

[14] Y. Galerne, A. M. Figueiredo Neto, and L. Liébert (1987). Microscopical structure of the uniaxial and biaxial lyotropic nematics. *J. Chem. Phys.*, 87: 1851–1856.

[15] P. Tolédano and A. M. Figueiredo Neto (1994). Topological metamorphosis in the phase diagram of a liquid of biaxial micellar aggregates. *Phys. Rev. Lett.*, 73: 2216–2219.

[16] Y. Hendrikx, J. Charvolin, and M. Rawiso (1986). Uniaxial-biaxial phase transition in lyotropic nematic solutions: Local biaxiality in the uniaxial phase *Phys. Rev. B*, 33: 3534–3537.

3
Experimental techniques

3.1 Introduction

In this chapter we will discuss some of the most employed experimental techniques in the study of the properties of lyotropic liquid crystals (LLC). We start with Polarized Optical Microscopy (POM) that is, probably, the one firstly used to observe the outstanding optical properties of these materials. Laser Conoscopy is, mainly, used to precisely measure optical birefringences and the nature of phase transitions in nematic lyotropic liquid crystals. Differential Scanning Calorimetry (DSC) is employed to investigate some of the LLC phase transitions as a function of temperature and the thermal behaviour of them. X-ray (SAXS) and neutron (SANS) diffraction and scattering techniques are employed to study structure and molecular nanosegregation in the basic units of the mesophases. Nuclear Magnetic Resonance (NMR) is a powerful and complementary tool to investigate molecular structure and dynamics in these materials. Finally, Cryogenic Transmission Electron Microscopy (cryo-TEM) for direct imaging is a technique that allows the observation of the specimen thermally fixed in its natural state.

An extensive bibliography is given at the end of this chapter, allowing readers to get more details about each technique.

3.2 Polarizing Optical Microscopy (POM)

The proposal and construction of the first polarized-light microscope is attributed to William Henry Fox Talbot [1] in 1834. Since then, this instrument has been shown to be very useful in different fields of research. It allows a first characterization of materials and the determination of some of their properties. Among these properties, birefringence is the most common. A material is birefringent when it presents the phenomenon of double refraction. Let us consider a uniaxial material. It has an axis, named the *optical axis*, along which the incident light beam propagation is governed by the index of refraction n_o, or the ordinary index of refraction. The light propagating along the optical axis has the polarization direction always perpendicular to it. When an unpolarized light beam reaches a transparent material along a direction different from that of the optical axis, the beam splits in two, with orthogonal polarization directions. One of them has the propagation governed by n_o and is named the ordinary ray, and the other, named the extraordinary ray, is governed by a direction-dependent refractive index. If the light beam direction is perpendicular to the optical axis, its propagation in the material is governed by the

Lyotropic Liquid Crystals. Ingo Dierking and Antônio Martins Figueiredo Neto, Oxford University Press.
 DOI: 10.1093/9780191924927.003.0003

index of refraction n_e, and is the extraordinary ray. Both rays when leaving the specimen experience an optical retardation, or path difference, defined as $\Gamma = d\Delta n$, where d and $\Delta n = |n_e - n_o|$ are the specimen thickness and birefringence, respectively. It is useful to consider Γ in terms of the wavelength (λ) of the light employed in the observation of the specimen. After passing through the specimen in the POM, both rays may interfere in an analyser, with the easy axis oriented perpendicularly to that of the polarizer placed before the specimen. If white light is employed, different colours and intensities are observed in the objective of the microscope.

Textures observed in POM (under crossed polarizers) of uniaxial lyotropic liquid crystals encapsulated in a particular sample holder inform about the topology of the director in this holder. Defects may be characterized and analysed as a function of external fields and temperature. An interesting example is the use of an external magnetic field in a uniaxial lyotropic ferronematic (lyotropic nematic doped with ferrofluids [2]) sample encapsuled in flat microslides to produce depletion layers [3,4]. The magnetic nanoparticles were shown to be trapped inside elliptical inversion walls, avoiding the complete relaxation of the defects generated in a previous planar sample, after the application of a uniform magnetic field perpendicularly to the original director orientation.

The analysis of the texture of different mesophases encapsulated in sample holders in the geometry of a slab allows the determination of important features of the mesophase. In particular, when ferrofluid doping is done in lyotropics, with small permanent magnets in the microscope stage, experimentalists are able to apply external magnetic fields of about 1 kG to the sample, follow the time-evolution of textures, and identify the topology of the mesophase optical axes. Typical textures are presented in Figure 3.1a–h.

However, due to a limitation of the equipment and the typical values of the optical birefringences of lyotropics (of the order of 10^{-5}), thick samples are required to access several interference fringes in a measurement of the birefringences, POM in the conoscopic layout is not adequate to this type of measurements. Qualitative information may be obtained using the POM with respect to the sign of the birefringence of uniaxial lyotropic liquid crystals and the birefringence [6,7], but with less accuracy. To perform reliable measurements of the optical birefringences and analyse phase transitions, another technique is necessary, laser conoscopy, that will be discussed in the next section.

3.3 Laser conoscopy

In 1976 Galerne and co-workers [8] proposed an elegant adaptation of the usual POM conoscopy technique to liquid crystalline samples with low birefringence, laser conoscopy. This delivers highly precise measurements, since a large number of interference fringes may be observed in the conoscopic interference pattern, which allows accurate studies of phase transitions in materials. It has been used not only with lyotropic [9] and thermotropic [10] liquid crystals, but also with large solid

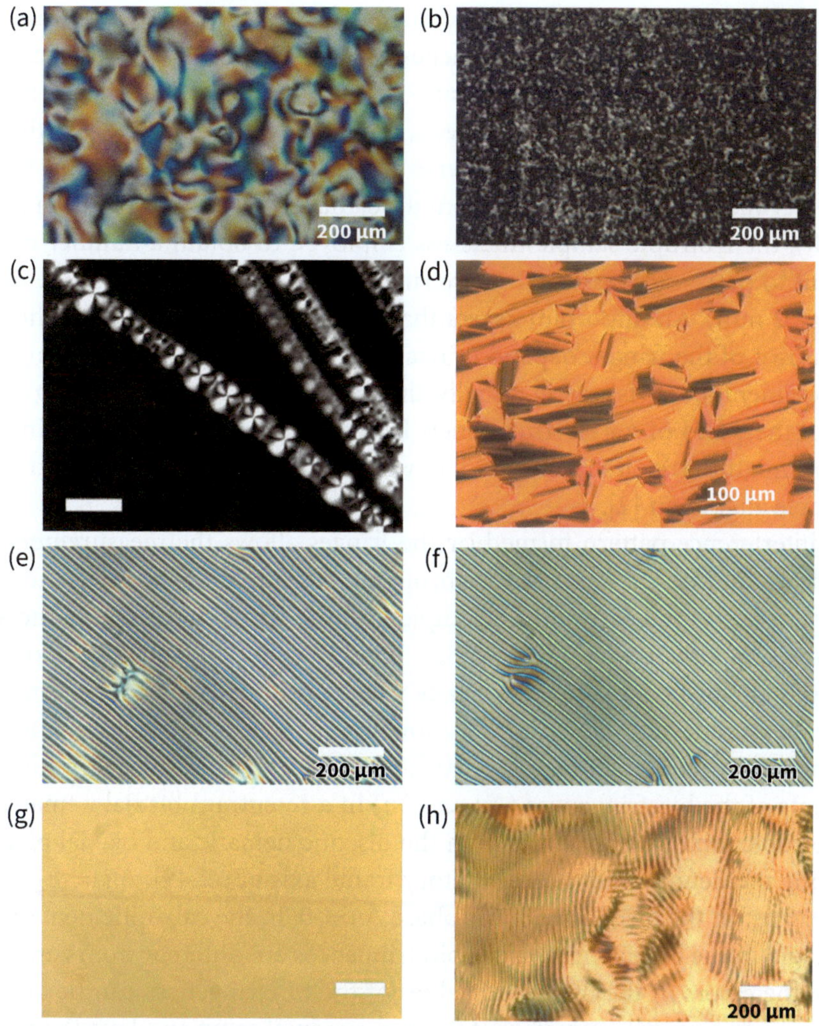

Figure 3.1 Lyotropic mesophase textures observed under crossed polarizers; the polarized optical microscopy (POM) technique. (a) Typical image of non-aligned nematic calamitic mesophase in a 200 μm thick glass microslide. Mixture of dodecyltrimethylammonium bromide/NaBr/1-dodecanol/water. T = 20 °C. (b) Typical image of non-aligned nematic discotic mesophase in a 200 μm thick glass microslide. Mixture of dodecyltrimethylammonium bromide/NaBr/1-dodecanol/water. T = 24 °C. (c) Typical image of lamellar mesophase. Mixture of cetylpiridinium chloride/hexanol/brine. White bar corresponds to 20 μm. Photo by Professor Dr. Oleg D. Lavrentovich, Université de Paris-Sud, Orsay, France. (d) Typical image of hexagonal mesophase. Hexagonal phase of LCLC sunset yellow. Photo by Miss Kelly Anton, laboratory of Professor Dr. Oleg D. Lavrentovich, Kent State University, USA. (e) Typical image of aligned cholesteric discotic mesophase in a 200 μm thick glass microslide. As the local director aligns perpendicular to the magnetic field, the cholesteric arrangement is maintained and the cholesteric axis aligns parallel to the external magnetic field. Mixture of potassium laurate/K_2SO_4/1-nonanol/water/brucine sulfate. T = 25 °C. (f) Typical image of aligned cholesteric biaxial mesophase in a 200 μm thick glass microslide. Also here, the cholesteric axis aligns parallel to the external magnetic field. Just the homeotropic regions (black stripes present in Figure 3.1(e)) are absent here. Mixture of potassium laurate/K_2SO_4/1-nonanol/water/brucine sulfate. T = 12.5 °C. (g) Typical image of magnetically aligned cholesteric

crystals [11]. Briefly, the optical setup consists of a polarized laser beam (e.g., a He-Ne laser of power ~15 mW) that reaches a beam expander used to increase the original diameter of the Gaussian beam. After the expander, the beam reaches a large-aperture microscope objective (e.g., 50× 0.85 P) that focuses it in the centre of the sample. A crossed linear polarizer is placed after the sample to allow the interference between the rays emerging from the sample, that are projected on a screen. Typically, the lyotropic sample thickness employed to obtain a number of fringes sufficient to precisely measure the birefringences is of the order of some millimetres (e.g., 2 mm). An important aspect that must be pointed out is that the quality and contrast of the fringes depend critically on the sample alignment throughout the experiment. For this reason, usually, the laser conoscopy optical setup is placed between the poles of an electromagnet. In some experiments, the doping of the lyotropic mixture with a small amount of water-based ferrofluids is used to improve the sample alignment [2].

The interference pattern formed by the fringes allows the measurement of the angles from the vertical axis passing through the centre of the sample and the one passing through the centre of the sample and the fringe position on the screen. In Figure 3.2a–d the typical conoscopic-fringe patterns of a lyotropic mixture with the three nematic phases are shown. The fringe positions and sample temperature may be measured with high accuracy, allowing an error in the determination of the birefringences of the order of $10^{-4} - 10^{-5}$. Figure 3.3 shows a typical result of measurements of the birefringences (Δn and δn) in a re-entrant biaxial nematic phase of a lyotropic mixture presenting both the discotic nematic and biaxial phases [9]. Following the definition of the laboratory frame axis of ref. [9], $\Delta n = n_2 - n_1$ and $\delta n = n_3 - n_2$. In the discotic nematic phase $\Delta n = 0$, in the calamitic nematic phase $\delta n = 0$, and in the biaxial phase both birefringences are different from zero.

Since the errors in the measured values of the birefringences can be very small (depending on several parameters of the experimental setup employed), these measurements can be used to investigate critical behaviour [12] of the uniaxial to biaxial phase transition, in particular deviations from the mean-field theory.

The conoscopy technique may be used for measurements of the tilt angle in uniaxial nematics oriented by surface effects (e.g., wetting) and fields. When the optical angle of the liquid crystal is tilted with respect to the propagation direction of the incident light beam, the interference pattern is dislocated in the plane of the screen. The precise determination of the fringe positions allows the measurement of the tilt angle with precision of about 2% [13].

Figure 3.1 (*Continued*) calamitic mesophase in a 200 μm thick glass microslide. As the local director aligns parallel to the magnet field, the cholesteric arrangement is destroyed. Mixture of potassium laurate/K_2SO_4/1-dodecanol/water/brucine sulfate. T = 25 °C. (h) Typical image of the cholesteric calamitic mesophase in a 200 μm thick glass microslide. Same sample of Figure 3.1(g), left to relax without any external magnetic field. The cholesteric arrangement is now visible. Mixture of potassium laurate/K_2SO_4/1-dodecanol/water/brucine sulfate. T = 25 °C.

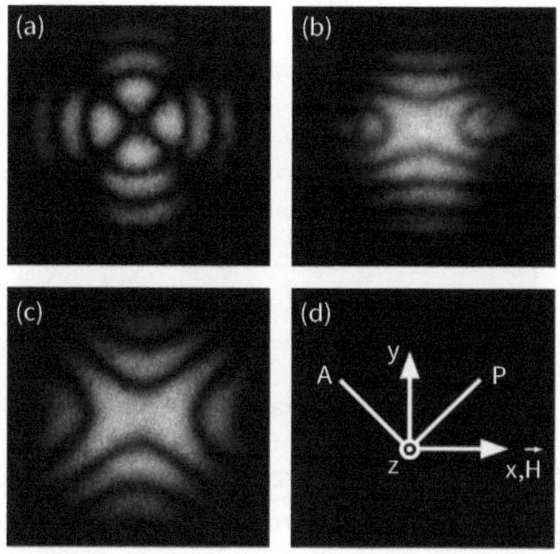

Figure 3.2 Conoscopic fringe patterns of a lyotropic mixture presenting the three nematic phases. Mixture of dodecyltrimethylammonium bromide/NaBr/1-dodecanol/water: (a) uniaxial discotic nematic phase with the director oriented perpendicular to the picture, and parallel to the laser beam propagation direction. T = 32 °C; (b) biaxial nematic phase. T = 30.1 °C; (c) uniaxial calamitic nematic phase with the director oriented in the plane of the picture, and perpendicular to the laser beam propagation direction. T = 26 °C; (d) geometry of the experiment, with the laboratory frame axes: P and A refer to the polarizer and analyser, $\vec{H}$ represents the external applied magnetic field.

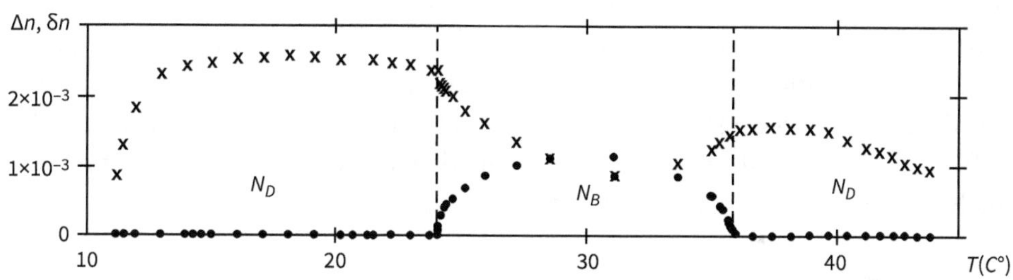

Figure 3.3 Optical birefringences $\Delta n = n_2 - n_2$ (•) and $\delta n = n_3 - n_2$ (x), as a function of temperature. N_D and N_B represent the uniaxial discotic and biaxial nematic phases.

3.4 Differential Scanning Calorimetry (DSC)

This technique measures the relative difference of energy given to a sample under analysis and a reference substance to maintain both at the same temperature, subjected to a controlled (pre-defined) temperature ramp. Two operation principles are employed in DSC equipment, one based on the power compensation and another on the heat flux. In the first one, sample and reference have individual heaters and the powers dissipated by the heaters are related to the process occurring in the sample. In the second one, both the sample and reference are placed on a metal disc that is heated (or cooled) and the temperature difference between them is measured. In terms of the physical-chemical parameters obtained, both principles are essentially equivalent.

DSC is widely used by experimentalists to characterize the nature of the phase transitions in thermotropic and lyotropic liquid crystals. It is a useful and

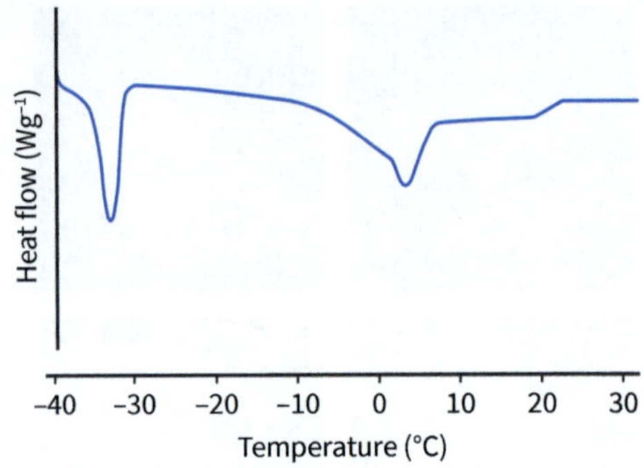

Figure 3.4 Thermogram of oleogels from glycerol-based lyotropic liquid crystals. The mixture is made of glycerol monooleate (GMO)/decane/water, at water:GMO weight ratios 3:1.

straightforward technique, allowing access to the thermal characteristics of the material, mainly in the vicinity of phase transitions, but not only there. DSC equipment is often present in laboratories that study liquid crystals, nano- and biological materials, and it is used as a complementary technique [14].

Figure 3.4 shows thermograms of oleogels from glycerol-based lyotropic liquid crystals [15]. Mixtures are made of glycerol monooleate (GMO)/decane/water. In Figure 3.4, the endothermic peak at about −33 °C corresponds to the decane melting. The broad peak at around 3 °C corresponds to the melting of glycerol in the mixture GMO/water. At around 18 °C there is also another endothermic event corresponding to the formation of a lamellar phase. At around 22 °C an exothermic peak indicates the transition to the reverse hexagonal phase. Obviously, the identification of the different liquid crystalline phases cannot be made only with the DSC results, but from the combination of results obtained with complementary techniques, namely, POM, diffraction, and scattering techniques [16,17].

3.5 X-ray and neutron diffraction and scattering

These techniques are employed to determine the structure and local ordering of the nanoscopic constituents of the lyotropic liquid crystals mesophases. The typical wavelength of the particles used in experiments is of the order of the Angstrom: $\lambda_x \sim 1$ Å; $\lambda_n \sim 5$ Å, for the X-ray and neutron, respectively. The range of the modulus of the reciprocal-space momentum-transfer vector $\vec{q}$, $q = 4\pi \sin(\theta)/\lambda$, where 2θ is the scattering angle is, typically, from 10^{-3}Å^{-1} to 1 Å^{-1}, allowing the investigation of structures between about 1 nm and 500 nm. The experimental resolution of SAXS is, in general, higher than in SANS. Both techniques are complementary and are used to give not only the structure and symmetry of the scatterers but also details about eventual molecular nanosegregation present in the basic units of lyotropics. We will come back to this point further on.

The normalized scattering intensity by a set of scatterers depends on a parameter named *scattering length*, *b*, that corresponds to the amplitude of the scattered wave, compared to the amplitude of the incident wave [18]. In the case of X-rays, *b* is proportional to the atomic number, *Z*, of the atom. In the case of neutrons, *b* presents a non-systematic dependence on the atomic number, being critically dependent on the particular isotope of the substance. The simple presence of an additional neutron in the atom nucleus may drastically change the scattering length of a scatterer (e.g., H_2O and D_2O). The (coherent) scattering length of the 1H is −3.74 fm and of 2H 6.67 fm [19].

In the 1960s, Luzzati and co-workers published seminal papers describing the structure of lyotropic systems, mainly lipid/water mixtures, using the X-ray diffraction technique and POM [20–22]. Diffraction peaks identified in their experiments allow the identification of lamellar (named neat soap), hexagonal (named middle soap), rectangular centred, and orthorhombic body-centred phases. Besides the symmetry of the different phases, structural parameters were obtained. This technique was also used to describe less positional ordered lyotropic mesophases, as micellar nematics [23,24].

Strictly speaking, the small-angle scattering would account for the broadening of the X-ray incident beam profile by meso-scale (generally 1–100 nm) scatterers. However, the term SAXS is commonly used also to describe the X-ray scattering profile curves (scattered intensity as a function of *q*) in the interval $\sim 0.005 < q < \sim 0.5\,\text{Å}^{-1}$.

Let us discuss in more detail the X-ray scattered intensity *I* as a function of *q* from lyotropic micellar systems. The profile of $I(q)$ is more complex than those from the usual micellar isotropic, lamellar, and hexagonal phases. The existence of short-range positional ordering promotes the presence of diffraction bands, superimposed to the usual small-angle scattering. The scattered intensity has the contribution from the form factor of the scatterers and the structure factor. The form factor depends on the geometric characteristics and the contrast of electronic density between the scatterer and the medium. The structure factor represents the spatial arrangement of the scatterers. Figure 3.5 shows a typical $I(q)$ curve of a lyotropic nematic liquid crystal.

In 2005, Förster and co-workers proposed a method to quantitatively fit the entire $I(q)$ curve from SAXS and SANS experimental results for a large number of nano- and mesoscopic materials [25]. The method allows the decoupling between the form and structural factors. Improvements on the method to take into account the resolution of the experimental setup were proposed in the following [26,27].

This advanced SAXS modelling was recently applied in the description of lyotropic uniaxial nematic liquid crystals [28]. The theoretical scattering intensity were written as:

$$I_{mod}(q) = SC \times S_{HS}(q) \times I_{mic}(q) \times S_{Lam}(q) + Back, \tag{3.1}$$

where $I_{mic}(q)$ is the intensity form factor of the polydisperse core-shell micelles; $S_{Lam}(q)$ is the pseudo-lamellar structure factor that describes the short-range positional ordering of the micelles; *SC* is an overall scale factor and *Back* a constant background. $S_{HS}(q)$ is the hard-sphere interaction structure factor, that will be discussed later. The versatility of the modelling allows the use of different structure and form factors to choose the most adequate to describe the scattering data. The solid lines in Figure 3.5 are best fits of Eq. 3.1 to the experimental SAXS scattered intensity of a magnetically oriented lyotropic nematic calamitic mesophase [28], showing the quality of the fitting procedure. Parameters from the form and structural factors are obtained.

As a working example of the application of this advanced analysis, we will discuss the scattered intensity from a uniaxial lyotropic nematic sample, illuminated by an X-ray beam, as a function of q [28].

A crude analysis, measuring the q-position of the peak of the 'Bragg-like diffraction band' and its width at half height, gives a good number for the repeating distance d (Bragg's equation – Eq. 3.2a) responsible for this band, and the correlation length D (Scherrer's equation – Eq. 3.2b).

$$\lambda = 2\, d\, sin\theta, \tag{3.2a}$$

$$D = \frac{\lambda}{\Delta(2\theta)\, cos\theta}, \tag{3.2b}$$

where $\Delta(2\theta)$ is the full-width at half height (FWHM) of the band, in radians. However, as pointed out before, a more detailed analysis allows the obtention of a complete set of parameters, besides d and D from a fit of the entire scattering curve.

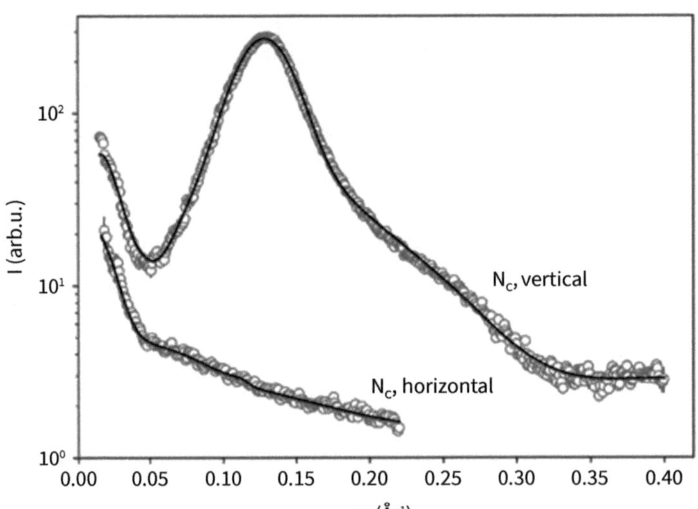

Figure 3.5 Typical scattered intensity *I* as a function of *q* from a lyotropic calamitic nematic phase, oriented by external magnetic field along the horizontal and vertical directions. Mixture of potassium laurate/K_2SO_4/dodecanol/water [28]. Open circles: Experimental data. Solid lines: Model fits.

(Reproduced with permission from ref. [28]).

Since the scattered intensity depends on the structure and form factors of the scatterers, a first step is to dilute the former nematic lyotropic mixture until the isotropic phase is reached. In this condition, we expect the predominance of the micellar structure factor in the scattered intensity ($I(q)$). In this case, $I(q)$ is written as:

$$I(q) = SC \times I_{mic}(q) + Back, \tag{3.3}$$

where

$$I_{mic}(q) = \int_0^{\infty} P_{mic}\left(q,\, R_1, t,\, \epsilon\right) D\left(R_1,\, \langle R_1 \rangle,\, \sigma_{R_1}\right) d\, R_1,$$

$$P_{mic}(q, R_1, t, \epsilon) = \int_0^{\pi/2} F_{cs}^2 \sin \alpha\ d\alpha,$$

P_{mic} is the form factor of the core-shell ellipsoidal micelle and $D\left(R_1,\ \langle R_1 \rangle,\ \sigma_{R_1}\right)$ is the radius distribution function:

$$D\left(R_1,\, \langle R_1 \rangle,\, z\right) = \left(\frac{z+1}{\langle R_1 \rangle}\right) \frac{R_1^z}{\Gamma\left[z+1\right]}\ exp\left(-\ (z+1)\ \frac{R_1}{\langle R_1 \rangle}\right),$$

$$r = R_1 \sqrt{(sin\ \alpha)^2 + \epsilon\ (cos\, \alpha)^2)}\,;$$

$$z = \frac{1}{\left(\frac{\sigma_{R_1}}{\langle R_1 \rangle}\right)^2}.$$

R_1, t, ϵ, σ_{R_1}, and α are the minor radius of the core of the micelle, the shell thickness, the ellipsoidal anisotropy (major radius $R_2\ = \epsilon\, R_1$), the micelle polydispersity, and the angle between the z axis and the ellipsoid axis, respectively. Γ is the Gamma function. The modulus of the core-shell scattering amplitude F_{CS} is given by:

$$F_{CS} = \frac{\Delta\rho_{shell}}{\Delta\rho_{core}} \left(V_{core} + V_{shell}\right)\, F_{sph}\ (q,\ r+t) \left(1 - \frac{\Delta\rho_{shell}}{\Delta\rho_{core}}\right)\ V_{core}\, F_{sph}\,(q, r)\,,$$

where, $\Delta\rho_{shell}$, $\Delta\rho_{core}$ are the electronic-density contrast between the shell and core, with respect to the solvent, V_{core} and V_{shell} the volumes of the core and shell, and F_{sph} is the modulus of the scattering amplitude of a sphere of radius r. This last function is necessary because, even at a high dilution (isotropic micellar phase), there is an interaction between micelles, and this function avoids one micelle occupying the volume already occupied by another. In the example discussed in ref. [28], the experimental scattered intensity at $q\ >\ 0.2\ \text{Å}^{-1}$ is fitted with Eq. 3.3, with the fitting parameters R_1, t and ϵ. So, the dimensions of the form factor of the micelles are obtained.

In a second step, the scattered intensity of the nematic mesophase, in an extended range of q (typically, $0.025\ <\ q\ <\ 0.4\ \text{Å}^{-1}$), can be fitted with the complete model

(Eq. 3.1), where the pseudo-lamellar structure factor (S_{Lam}) is incorporated:

$$S_{Lam} = 1 + \beta(q)\,(Z_0(q) - 1)\;G(q),$$

where $\beta(q) = \langle F_{CS}(q)\rangle^2 / \langle F_{CS}^2(q)\rangle$, and

$$Z_0(q) = \frac{c(2\pi)^{d-1}}{n v_d \Omega q^{(d-1)}} \sum\nolimits_{hkl} m_{hkl} f_{hkl}^2 L_{hkl},$$

where for the pseudo-lamellar structure (periodicity a), $m_{h00} = 1$; $d = 1$; $f_{h00} = 1$; $n = 1$; $v_d = a$; $\Omega = 1$ and $q_h = 2\pi h/a$. $L_{h00} = \{\delta/2\pi\}/\{q^2 + (\delta/2)^2\}$ is the peak-shape function, $G(q) = \exp\left[-\sigma_a^2 a^2 q^2\right]$ is the Debye–Waller factor, and σ_a the Debye–Waller mean displacement. c introduces a weight for the Porod invariant for the pseudo-lamellar structure, and δ is the band width which is related to the domain size $D = 2\pi/\delta$. The Bragg-like band observed in the diffraction pattern of the lyotropic nematic shows a small asymmetry. This indicates an interaction between micelles arranged in the pseudo-lamellar ordering. To account for this effect, a hard-sphere interaction structure factor S_{HS} is introduced in Eq. 3.1:

$$S_{HS} = \frac{1}{1 + 24\eta\; G[2R_{HS}q]/2R_{HS}q},$$

$$G(A) = \frac{\alpha(\sin A - A\cos A)}{A^2} + \frac{\beta\left(2A\,sinA + \left(2 - A^2\right)cosA - 2\right)}{A^3} + \frac{\gamma\left\{-A^4\,cosA + 4\left(\left(3A^2 - 6\right)cosA + \left(A^3 - 6A\right)sinA + 6\right)\right\}}{A^5},$$

where $\eta = 8\sigma_a^2$, η_{HS} is the hard-sphere volume fraction,

$$\alpha = \frac{(1 + 2\eta)^2}{(1 - \eta)^4},\quad \beta = \frac{-6\eta\left(1 + \frac{\eta}{2}\right)^2}{(1 - \eta)^4},\quad \gamma = \frac{\eta\alpha}{2}.$$

From the fit of Eq. 3.1 to the experimental data, the parameters a and D, the structure periodicity and correlation length, respectively, in different directions of the SAXS pattern of an oriented sample can be obtained.

Besides these parameters, SAXS allows the measurement of the nth -order orientational order parameter $\langle P_n(cos\phi)\rangle$, where $P_n(cos\phi)$ represents the Legendre polynomial functions [29] and ϕ the angle between the director and orientation (symmetry) axis of a particular micelle. The symbol $\langle \ldots \rangle$ denotes an averaging over all the micelles. To measure the order parameter a two-dimensional diffractogram in the plane of the reciprocal space momentum transfer vector (Figure 3.6) with an oriented sample has to be done. Due to the orientation distribution function of the local micellar symmetry axis along the director, the first-order diffraction band

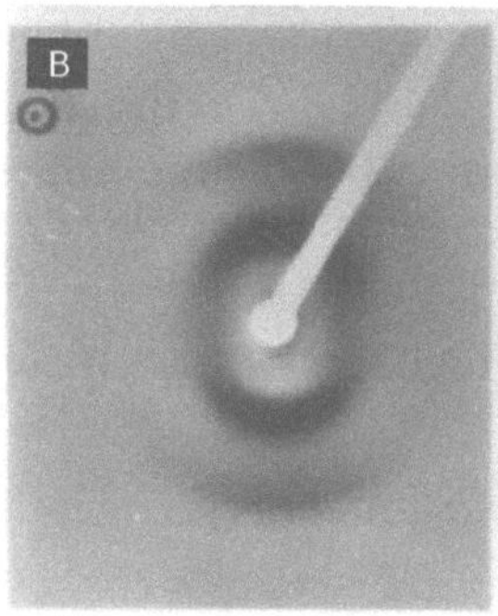

Figure 3.6 Typical 2D diffraction pattern of an oriented biaxial nematic mesophase. Mixture of potassium laurate/decanol/water. B represents the direction of the applied magnetic field.

intensity spreads around a circumference centred in the direction of the incident X-ray beam. The orientation distribution function can be expanded in a series of Legendre polynomial functions and the order parameters can be calculated from SAXS data [30]. The physical interpretation of the different $\langle P_n \rangle$ is associated with the degree of orientational ordering of the micelle's symmetry axis with respect to the director and the shape of the orientation distribution functions [31].

Differently from (and complementary to) the SAXS, the SANS allows experimentalists to use the contrast technique to investigate details in the nanoscopic structure of the basic units of lyotropic liquid crystals.

SANS was used to study the nanoscopic structure of lyotropic nematic liquid crystals, employing the contrast variation with deuterated molecules in the mixture composition [32,33]. The mixture investigated was the classical potassium laurate/decanol/water. The neutron scattering–length contrast variation is achieved using protonated potassium laurate and decanol, or perdeuterated laurate and protonated decanol, in various mixtures of heavy and light water. These results [33] indicated that the molecules of laurate and decanol are not uniformly distributed in the micelle, showing a local nanosegregation. Moreover, it showed that even in the uniaxial nematic phase, micelles are intrinsically biaxial, supporting the IBM model (see Chapter 5).

Neutron scattering and diffraction has been used to determine the orientational order parameters of lamellar and inverted hexagonal lipid-based lyotropic mesophases, doped with magnetic nanoparticles [34].

A remarkable and didactic example of the investigation of the nanostructure of scatterers in lyotropics, formed by triphilic star-polyphiles, using the contrast technique was reported by de Campo et al. [35]. From the peak positions obtained by SAXS, a hexagonal mesophase is identified. Also, SANS furnishes diffraction peaks consistent with the hexagonal phase, but these peaks are broader than those in SAXS due to instrumental resolution. To obtain the nanoscopic structure of the scatterers,

SANS experiments with different ratios of H_2O/D_2O wt/wt were carried out. Molecular nanosegregation details were revealed in the contrast experiments, allowing the proposition of the local arrangement of the molecules in the hexagonal structure.

3.6 Nuclear Magnetic Resonance (NMR)

NMR is widely used in lyotropic liquid crystals research as a complementary technique to SAXS, SANS, and POM [36–38]. The study of the line shape and relaxation behaviour of atoms present in the molecules that compose the lyotropic mixtures (e.g., ^{2}H, ^{13}C, and ^{17}O) allowed the investigation not only of the structure of lyotropics, but also, in some cases [38], the behaviour of their basic units in the presence of an external magnetic field. The local curvature of lyotropic structures and characteristic length scales of the molecular arrangements can be obtained from relaxation behaviour analysis through NMR measurements. In the case of D_2O, the electric field gradient (EFG) at the deuteron has cylindrical symmetry around the bond axis, and the quadrupole splitting is proportional to the degree of order of the bond axis.

The line shape analysis informs about the symmetry of the structure, allowing the identification of uniaxial and biaxial phases, cubic and rectangular molecular arrangements. The analysis of the D_2O quadrupole splitting in ternary lyotropic mixtures of potassium laurate (KL)/decanol (DeOH)/water allowed the identification of the lyotropic nematic mesophases, in particular, the biaxial one [39].

Complementary information about the structure and molecular arrangement in lyotropics can be obtained through the measurement of relaxation rates (time-correlation functions) from samples subjected to external magnetic fields. The lab-frame spectral densities measured in NMR experiments can be used to construct the crystal-frame spectral densities, that characterizes the *Crystallographic Point Group* of the lyotropic mesophase [40].

In conventional NMR experiments, samples are subjected to a time-dependent magnetic induction (radio-frequency field) tuned in resonance with the Larmor frequency (υ_L) of the nuclei under investigation. Samples are subjected to a constant (Zeeman) magnetic induction [41]. As the Larmor frequency is proportional to the Zeeman applied magnetic induction, situations where this field is very small practically makes the use of this technique unfeasible. To overcome this problem *Field-cycling Nuclear Magnetic Resonance (FC-NMR)* was proposed and has been widely used since then. The basis of the technique is allowing the Zeeman magnetic induction to vary as a function of time (cycled field). In these conditions, in principle, there is no limitation to the value of the Zeeman induction [42].

Rommel and co-workers used the FC-NMR to study the proton spin relaxation dispersion (T_1 – spin-lattice relaxation) of phospholipid membranes [43]. They found that the proton relaxation phenomenon, due to the orientational ordering fluctuations, is present in the kilohertz regime. The reorientation of individual molecules contributes in the high-frequency regime. Rodrigues and co-workers [44] investigated the ^{1}H spin-lattice relaxation rate in a lyotropic uniaxial nematic mesophase (mixture of KL/DeOH/H_2O/D_2O), over a range of four orders of magnitude in frequency. The measurements T_1 (υ_L) at different temperatures were performed in a home-built field-cycling NMR apparatus [11], in the υ_L range from 2×10^3 *Hz* to 6.6 *MHz* (Figure 3.7). The mechanisms proposed to describe the experimental results are: molecular reorientation by translational diffusion of KL and DeOH molecules (see, e.g., [45] for more details about translational molecular diffusion); molecular exchange between the micelle and the bulk; and order fluctuations of the local micellar director.

More recently, NMR was employed as a complementary technique to investigate aspects of the phase behaviour, self-assembling, and tautomeric equilibrium in sunset yellow (SSY)/water lyotropic mixtures [46]. The explored resonance frequencies were 400 and 40 MHz for ^{1}H and ^{15}N. The mixtures exhibit the isotropic, nematic, and columnar phases. It was shown that, in the process of growing the SSY aggregates, water is segregated from the intercolumnar areas. Moreover, the lifetime of the SSY aggregates was estimated to be larger than $\sim10^{-8}$ *s*.

As a final remark, the NMR technique remains a powerful and complementary tool to investigate molecular structure and dynamics in materials. Important improvements have been proposed in the technique and more are expected to come.

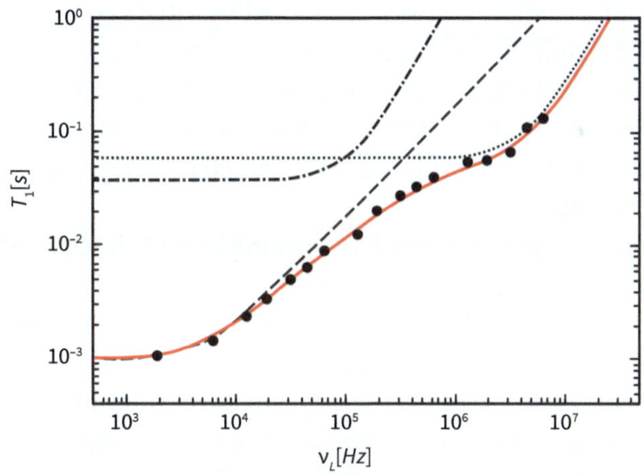

Figure 3.7 T_1 (υ_L) experimental results from the lyotropic nematic calamitic mesophase at 30 °C. Mixture of potassium laurate (KL)/decanol (DeOH)/H_2O/D_2O. (i) Dot, (ii) dash dot, and (iii) dash are contributions from: (i) molecular reorientation by translational diffusion of KL and DeOH molecules, (ii) molecular exchange between the micelle and the bulk, and (iii) order fluctuations of the local micellar director, respectively. The solid line is the fit using all the contributions.

3.7 Cryo-TEM for direct imaging

Differently from the usual Transmission Electron Microscopy (TEM), the cryogenic TEM (cryo-TEM) is a technique that allows the observation of the specimen thermally fixed in its natural state [47,48]. Typically, a stable thin layer, smaller than 3×10^3 Å, is formed when a liquid is spread over a grid. After that, the specimen is cooled and transferred to the TEM, without warming above the devitrification temperature (~140 *K*) [48].

One of the first experiments employing the Cryo-TEM with lyotropic mixtures (aqueous solutions of cetyltrimethylammonium chloride, with salicylate as counterion) was reported in 1992 [49]. Images show the transition from globular (typically 5 *nm* in diameter) to threadlike micelles in samples with different relative concentrations of the mixture components. The modifications of the shape of the molecular aggregates were correlated to the rheological behaviour of the sample.

Spherical and cylindrical micelles were observed by cryo-TEM [48] in lyotropic mixtures of cetyltrimethylammonium bromide (CTAB) in glycerol and sodium dodecyl sulfate (SDS) in formamide. Samples of the mixtures, placed in the grid, were plunged firstly into liquid ethane, and then into liquid nitrogen for the TEM observation. No damage provoked by the electron beam was detected in the observations.

The cryo-fracture transmission electron microscopy (cryo-fracture TEM) is an alternative method to observe nanostructures in lyotropics [50]. The technique consists of the preparation of a thick film of the sample formed on the carbon grid. The temperature of the specimen is lowered below vitrification and then the grid is punched with a needle to create perforations. The edges of the specimen are, then, observed in an electron microscope. Lyotropic mixtures of octaethylene glycol mono-n-dodecyl ether ($C_{12}E_8$)/water and Pluronic P84 (EO_{19}–PO_{43}–EO_{19})/water were investigated with this technique and the lamellar, hexagonal, and cubic structures were clearly imaged.

Gao and co-workers [51] reported a comprehensive description of specimen preparation methods to enable high-resolution direct cryo-TEM and replica freeze-fracture TEM observations of lyotropic and thermotropic liquid crystals. Their experiments show subnanometre-resolution cryo-TEM observations of the molecular packing of columnar aggregates in lyotropic chromonic LCs. To visualize the (vitrified) bulk structure of the liquid crystalline sample, the technique of Cryo-Electron Microscopy of Vitreous Section (CEMOVIS), where high-pressure freezing is followed by cryo-ultramicrotomy, was employed. Figure 3.8 [51] shows the CEMOVIS images of nematic regions in 15% disodium cromoglycate (DSCG)/water/dextran.

One of the most challenging collateral effects of the use of the cryo-TEM in studying soft matter is eventual electron beam-damage to the structure researchers are interested in investigating. Much progress has been achieved up to now not only to overcome this drawback, but also in hardware and software development,

Figure 3.8 Cryo-electron microscopy of vitreous section (CEMOVIS) image of nematic region of the lyotropic mixture of disodium cromoglycate (DSCG)/water/dextran.

better resolution, and enhanced contrast imaging. Advanced cryo-tomography will certainly provide 3D images of a sample's structure that will allow experimentalists to investigate not only the static aspect of them, but also the dynamics present in the formation of the nanoscale objects in lyotropics and living systems [52].

References

[1] H. F. Talbot Esq. M.P. F.R.S. (1834). XLIV. *Experiments on light, The London, Edinburgh, and Dublin Philosophical Magazine and Journal of Science*, 5:29: 321–334. doi:10.1080/14786443408648474.

[2] C. Scherer and A. M. Figueiredo Neto (2005). Ferrofluids: properties and applications. *Braz. J. Phys.*, 35; 718–727.

[3] L. Liébert and A. M. Figueiredo Neto (1984). Optical microscopic observations of depletion layers in a calamitic ferronematic lyomesophase. *J. Phys. (Paris) Lett.*, 45, L-173–178.

[4] A. M. Figueiredo Neto, Ph. Martinot-Lagarde, and G. Durand (1984). Anisotropy induced director refraction inside inversion walls in nematic liquid crystals. *J. Phys. (Paris) Lett.*, 45, L-793–798.

[5] A. M. Figueiredo Neto and S. R. A. Salinas (2005). *The Physics of Lyotropic Liquid Crystals, Phase Transitions and Structural Properties.* Oxford: Oxford University Press.

[6] A. E. Yavuz (2021). Determination of order parameter and birefringence depending on temperature of nematic calamitic mesophase formed by TTAB + water binary system using polarized microscopy. *Cumhuriyet Sci. J.*, 42, 152–155.

[7] D. A. Oliveira, D. D. Luders, G. P. Souza, N. M. Kimura, and A. J. Palangana (2009). An optical microscopy study of the biaxial – calamitic nematic lyotropic phase transition. *Cryst. Res. Technol.*, 44, 1255–1258.

[8] Y. Galerne, S. T. Lagerwall, and L. W. Smith (1976). Biaxialite critique d'un smectique C. *Optics Communications*, 19, 147–149.

[9] Y. Galerne and J. P. Marcerou (1985). Temperature-concentration behaviour of the order parameter in the nematic phases of a lyotropic liquid crystal. *J. Phys. (Paris)*, 46, 589–594.

[10] Y. Galerne (1978). Refractive index measurements at a 2nd order smectic A to C phase transition. *J. Phys. (Paris)*, 39, 1311–1316.

[11] A. I. Kolesnikov, R. M. Grechishkin, S. A. Tretiakov, V. Ya. Molchanov, A. I. Ivanova, E. I. Kaplunova, and E. Yu. Vorontsova (2013). Laser conoscopy of large-sized optical crystals. IOP Conf. *Series: Materials Science and Engineering*, 49, 012037.

[12] P. Boonbrahm and A. Saupe (1984). Critical behavior of uniaxial–biaxial nematic phase transitions in amphiphilic systems. *J. Chem. Phys.*, 81, 2076–2081.

[13] B. L. Van Horn and H. H. Winter (2001). Analysis of the conoscopic measurement for uniaxial liquid-crystal tilt angles. *Applied Optics*, 40, 2089–2094.

[14] P. Gill, T. T. Moghadam, and B. Ranjbar (2010). Differential scanning calorimetry techniques: Applications in biology and nanoscience. *J. Biomol. Tech.*, 21, 167–193.

[15] Y. Cegla-Nemirovsky, A. Aserin, and N. Garti (2015). Oleogels from glycerol-based lyotropic liquid crystals: phase diagrams and structural characterization. *J. Am. Oil Chem. Soc.*, 92, 439–447.

[16] C. Li, J. He, J. Liu, Z. Yu, Q. Zhang, C. He, and W. Hong (2010). Self-assembly of lyotropic liquid crystal phases in ternary systems of 1,2-dimethyl-3-hexadecylimidazolium bromide/1-decanol/water. *J. Colloid Interface Sci.*, 342, 354–360.

[17] J. Mahadeva, T. N. Govindaiah, R. Somashekar, and N. G. Malleshi (2009). Lyotropic liquid crystalline phases in the mixtures of non-mesogenic compounds. *Mol. Cryst. Liq. Cryst.*, 509, 21–29/[763–771].

[18] P. Schurtenberger (2002). Contrast and contrast variation in neutron, x-ray and light scattering. In: P. Lindner and Th. Zemb (eds), *Neutrons, X-Rays and Light*, Dordrecht: Elsevier, chapter 7.

[19] NIST Center for Neutron Research (2021). https://www.nist.gov/ncnr/neutron-scattering-lengths-list

[20] V. Luzzati, H. Mustacchi, A. Skoulios, and F. Husson (1960). La structure des colloïdes d'association. I. Les phases liquide–cristallines des systèmes amphiphile–eau. *Acta Cryst.*, 13, 660–667.

[21] F. Husson, H. Mustacchi, and V. Luzzati (1960). La structure des colloïdes d'association. II. Description des phases liquide–cristallines de plusieurs systèmes amphiphile–eau: Amphiphiles anioniques, cationiques, non-ioniques. *Acta Cryst.*, 13, 668–677.

[22] V. Luzzati (1968). X-ray Diffraction Studies of Lipid-Water Systems. In: D. Chapman (ed.), *Biological Membranes Physical Fact and Function*, chapter 3, Academic Press: London, pp 71–123.

[23] J. Charvolin, A. M. Levelut, and E. T. Samulski (1979). Lyotropic nematics: molecular aggregation and susceptibilities. *J. Phys. (Paris) Lett.*, 40, L-587–592.

[24] A. M. Figueiredo Neto, Y. Galerne, A. M. Levelut, and L. Liébert (1985). Pseudo-lamellar ordering in uniaxial and biaxial lyotropic nematics: a synchrotron x-ray diffraction experiment. *J. Phys. (Paris) Lett.*, 46, L-499–505.

[25] S. Förster, A. Timmann, M. Konrad, C. Schellbach, A. Meyer, S. S. Funari, P. Mulvaney, and R. Knott (2005). Scattering curves of ordered mesoscopic materials. *J. Phys. Chem.*, 109, 1347–1360.

[26] A. Sundblom, C. L. P. Oliveira, A. E. C. Palmqvist, and J. S. Pedersen (2009). Modeling in situ small-angle X-ray scattering measurements following the formation of mesostructured silica. *J. Phys. Chem. C* -, 113, 7706–7713.

[27] C. L. P. Oliveira, B. B. Gerbelli, E. R. T. Silva, F. Nallet, L. Navailles, E. A. Oliveira, and J. S. Pedersen (2012). Gaussian deconvolution: a useful method for a form-free modeling of scattering data from mono- and multilayered planar systems *J. App. Cryst.*, 45, 1278–1286.

[28] O. R. Santos, D. Reis, A. G. Oliveira-Filho, C. L. P. Oliveira, and A. M. Figueiredo Neto (2022). Structure and local order of lyotropic cholesteric calamitic phases: The effect of the chiral molecule. *J. Mol. Liq.*, 349, 118097.

[29] G. Luckhurst and R. N. J. Yeates (1976). Orientational order of a spin probe dissolved in nematic liquid crystals. An electron resonance investigation. *Chem. Soc., Faraday Trans. 2*, 72, 996–1009.

[30] M. Deutsch (1991). Orientational order determination in liquid crystals by x-ray diffraction. *Phys. Rev. A*, 44, 8264–8270.

[31] T. S. Yankova, A. Yu. Bobrovsky, and A. Kh. Vorobiev (2012). Order parameters ⟨P2⟩, ⟨P4⟩, and ⟨P6⟩ of aligned nematic liquid-crystalline polymer as determined by numerical simulation of electron paramagnetic resonance spectra. *J. Phys. Chem. B*, 116, 6010–6016.

[32] Y. Hendrikx, J. Charvolin, M. Rawiso, L. Liebert, and M. C. Holmes (1983). Anisotropic aggregates of amphiphilic molecules in lyotropic nematic phases. *J. Phys. Chem.* 87, 3991–3999.

[33] Y. Hendrikx, J. Charvolin, and M. Rawiso (1986). Uniaxial-biaxial phase transition in lyotropic nematic solutions: Local biaxiality in the uniaxial phase. *Phys. Rev. B*, 33, 3534–3537.

[34] J. J. Vallooran, S. Bolisetty, and R. Mezzenga (2011). Macroscopic alignment of lyotropic liquid crystals using magnetic nanoparticles. *Adv. Mater.*, 23, 3932–3937.

[35] L. de Campo, T. Varslot, M. J. Moghaddam, J. J. K. Kirkensgaard, K. Mortensen, and S. T. Hyde (2011). A novel lyotropic liquid crystal formed by triphilic star-polyphiles: hydrophilic/oleophilic/fluorophilic rods arranged in a 12.6.4. tiling. *Phys. Chem. Chem. Phys.*, 13, 3139–3152.

[36] B. Halle and I. Furó (1998). In: P. Tolédano and A. M. Figueiredo Neto (eds), *Phase Transitions in Complex Fluids*, Singapore: World Scientific, chapter 4.

[37] F. Y. Fujiwara, L. W. Reeves, M. Suzuki, and J. A. Vanin (1979). Studies of Lyotropic Liquid Crystals that Align in Magnetic Fields. In: K. L. Mittal (ed.), *Solution Chemistry of Surfactants*, Volume 1, New York: Plenum Press, 1, pp 63–77.

[38] B. J. Forrest and L. W. Reeves (1981). New lyotropic liquid crystals composed of finite nonspherical micelles. *Chem. Rev.*, 81, 1–14.

[39] L. J. Yu and A. Saupe (1981). Observation of a biaxial nematic phase in potassium laurate-1-decanol-water mixtures. *Phys. Rev. Lett.*, 45, 1000–1003.

[40] S. Gustafsson and B. Halle (1993). Group theoretical analysis of nuclear spin relaxation in liquid crystals and molecular solids. *Molec. Phys.*, 80, 549–582.

[41] F. Noack (1986). NMR field-cycling spectroscopy: principles and applications. *Prog. Nucl. Magn. Reson. Spectrosc.*, 18, 171–276.

[42] F. Noack, M. Notter, and W. Weiss (1988). Relaxation dispersion and zero-field spectroscopy of thermotropic and lyotropic liquid crystals by fast field-cycling N.M.R. *Liquid Crystals*, 3, 907–925.

[43] E. Rommel, F. Noack, P. Meier, and G. Kothe (1988). Proton spin relaxation dispersion studies of phospholipid membranes. *J. Phys. Chem.*, 92, 2981–2987.

[44] C. R. Rodríguez, F. Vaca Chávez, D. J. Pusiol, and A. M. Figueiredo Neto (2001). NMR study of the slow dynamics and local structure in lyotropic micellar liquid crystals. *Granular Matter*, 3, 53–56.

[45] G. Lindblom and G. Orädd (1994). NMR studies of translational diffusion in lyotropic liquid crystals and lipid membranes. *Prog. Nucl. Magn. Reson. Spectrosc.*, 26, 483–515.

[46] K. Kristinaitytė, A. Maršalka, L. Dagys, K. Aidas, I. Doroshenko, Y. Vaskivskyi, Y. Chernolevska, V. Pogorelov, N. R. Valevičienė, and V. Balevicius (2018). NMR, Raman, and DFT study of lyotropic chromonic liquid crystals of biomedical interest: Tautomeric equilibrium and slow self-assembling in sunset yellow aqueous solutions. *J. Phys. Chem. B*, 122, 3047–3055.

[47] M. Adrian, J. Dubochet, J. Lepault, and A. W. McDowall (1984). Cryo-electron microscopy of viruses. *Nature*, 308, 32–36.

[48] Z. Lin, H. T. Davis, and L. E. Scriven (1996). Cryogenic electron microscopy of micelles and lyotropic liquid crystals in some polar solvents. *Langmuir*, 12, 5489–5493.

[49] T. M. Clausen, P. K. Vinson, J. R. Minter, H. T. Davis, Y. Talmon, and W. G. Miller (1992). Viscoelastic micellar solutions: microscopy and rheology. *J. Phys. Chem.*, 96, 474–484.

[50] A. González-Pérez and U. Olsson (2008). Cryo-fracture TEM: direct imaging of viscous simples. *Soft Matter*, 4, 1625–1629.

[51] M. Gao, Y.-K. Kim, C. Zhang, V. Borshch, S. Zhou, H.-S. Park, A. Jákli, O. D. Lavrentovich, M.-G. Tamba, A. Kohlmeier, G. H. Mehl, W. Weissflog, D. Studer, B. Zuber, H. Gnägi, and F. Lin (2014). Direct observation of liquid crystals using cryo-TEM: Specimen preparation and low-dose imaging. *Microscopy Research and Technique*, 77, 754–772.

[52] A. Doerr (2017). Cryo-electron tomography. *Nature Methods*, 14, 34.

4
Computer simulations of liquid crystals

4.1 Introduction

Over the last three decades, since about 1990, computer simulations have developed into a serious tool to predict the structure and properties of materials. Soft-matter materials and liquid crystals have been no exception to this [1]. With a drastic increase in memory and computing power, as well as the introduction of parallel computing, the complexity and size of simulated systems have increased to a reasonable sample size to predict phases and thermodynamic properties of liquid crystals. Nevertheless, it should be noted that it is still a huge task to predict phase sequence and transition temperatures through atomistic simulations, simply because soft-matter systems are extremely sensitive to even small changes in molecular structure and environment.

The field of liquid crystal simulations strongly took off in the beginning of the 1990s and has since steadily increased, as can be seen by the number of published papers per year shown in Figure 4.1. Most of these simulations are concerned with thermotropic liquid crystals and their properties [2,3], due to their prominent position as materials in flat-panel displays. Interestingly, the area of lyotropic liquid crystals only contributes to about 2% of the published work. This is due to the difficulty in simulating lyotropics. Atomistic simulations of lyotropic liquid crystals are particularly complex [4], because of the additional solvent molecules needed, and especially the processes of micelle formation and subsequent aggregation of micelles into liquid crystalline structures, which takes place on a timescale very challenging for such simulations [5]. For example, the exchange of a single amphiphilic molecule from an existing micelle lies in the ns to μs range while the time required for the formation or destruction of whole micelles is of the order of ms all the way to seconds, by far too long for atomistic simulations.

In general, there are several different approaches to computer simulations of liquid crystals, depending on the length and timescale of the problem, as summarized in Figure 4.2. One can roughly place these into three different categories: (i) the microscopic scale in the region of nanometres; (ii) the mesoscopic range of micrometre length scale; and (iii) the macroscopic systems which are approximately of a millimetre size.

In the microscopic regime single molecules and interacting molecular structures are described, including bulk properties such as phase transition temperatures via order-parameter determination. Here, quantum mechanical and atomistic models are utilized, which cover a timescale of picoseconds and nanoseconds, respectively.

Lyotropic Liquid Crystals. Ingo Dierking and Antônio Martins Figueiredo Neto, Oxford University Press.
 DOI: 10.1093/9780191924927.003.0004

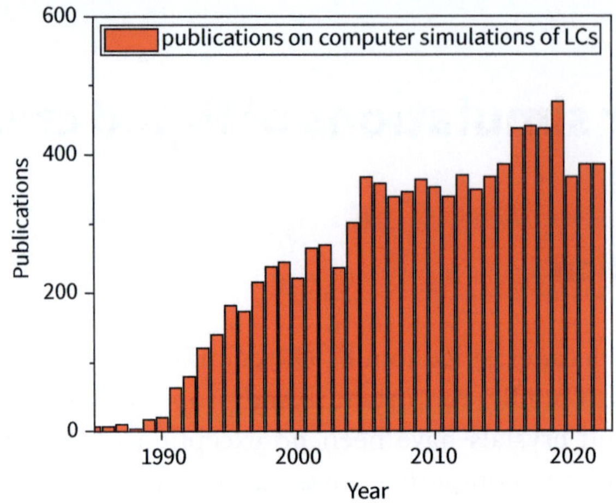

Figure 4.1 Number of publications related to computer simulations of liquid crystals over the last 40 years. Most of these simulations are related to thermotropic rather than lyotropic LC systems.

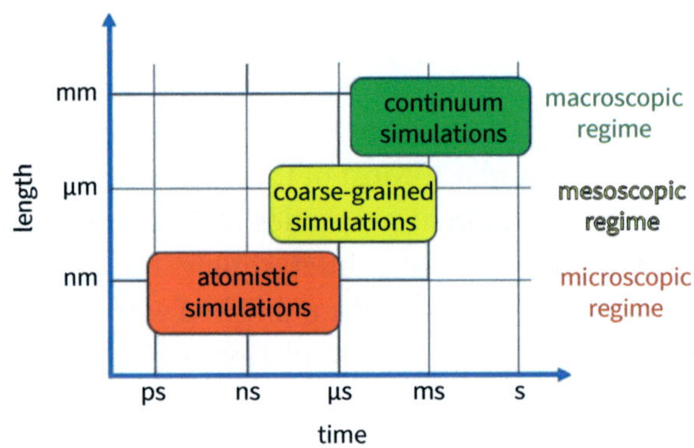

Figure 4.2 Different approaches that are used for computer simulations of liquid crystals depending on the spatial and temporal scale of the problem studied.

An illustrative example is shown in Figure 4.3 for the time development of the order parameter $S_2 = 1/2 \langle 3cos^2\vartheta_i - 1 \rangle$ when passing the isotropic to nematic phase transition, calculated from atomistic computer simulations [6].

Such atomistic simulations would require too large a number of molecules and too much atomic and molecular detail, thus too much computational power, to be carried out in the mesoscopic regime where the structure of more complicated phases or even liquid crystal polymers is simulated. Here, one has to resort to coarse-grained simulations where atomistic details are substituted by spheres, rods, or ellipsoids [6]. Generally, for these kind of simulations, several thousand approximated molecules are simulated in a box with continuous boundary conditions. An illustrative example is depicted in Figure 4.4 where mesogens are approximated by

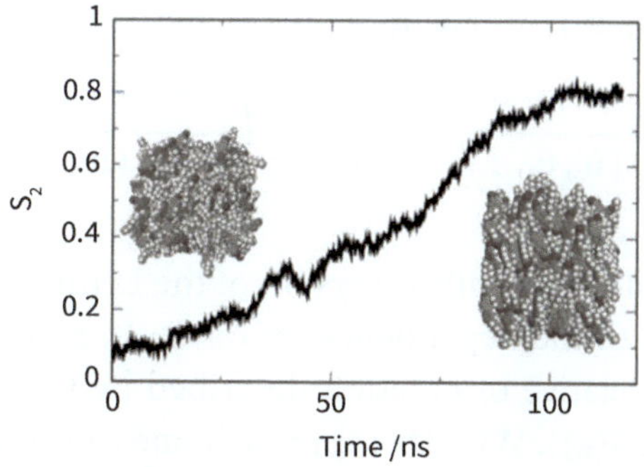

Figure 4.3 Example of an atomistic simulation of the transition from the disordered isotropic phase with an order parameter S≈0 to the orientationally ordered nematic phase with S≈0.8. Simulations of this kind are generally performed only in the microscopic regime with up to 512 molecules in a box with periodic boundary conditions and over a timescale of several hundreds of nanoseconds or less.
(Reproduced with permission from ref. [6].)

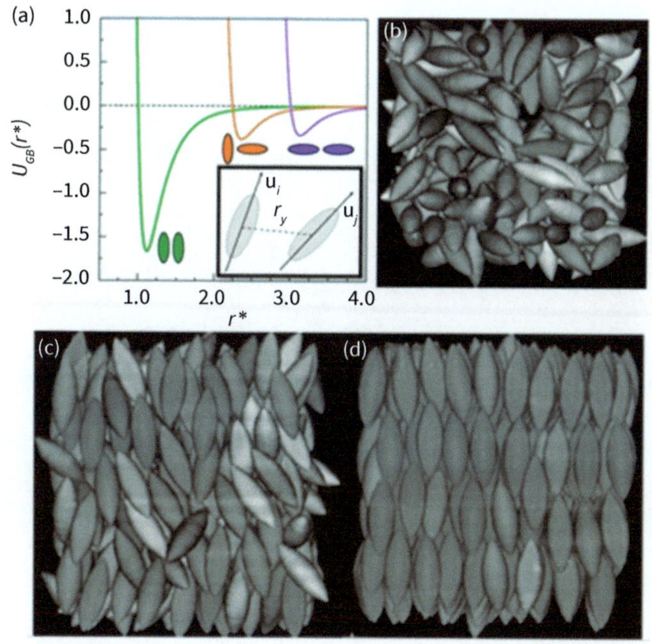

Figure 4.4 (a) The Gay–Berne potential of Eq. 4.1 plotted as a function of distance and orientation of two neighbouring molecules (see inset). Snapshots during the simulation of (b) the isotropic, (c) the nematic, and (d) the smectic phase.
(Reproduced with permission from ref. [6].)

soft ellipsoids whose interactions are described by a potential, in this case the very successful Gay–Berne potential, which allows the description of phase transitions by a change of temperature [7].

$$U_{ij} = 4\varepsilon_o \left[\varepsilon\left(\hat{\mathbf{u}}_i, \hat{\mathbf{u}}_j\right)\right]^{\nu}\left[\varepsilon'\left(\hat{\mathbf{u}}_i, \hat{\mathbf{u}}_j, \hat{\mathbf{r}}_{ij}\right)\right]^{\mu} \left[\left(\frac{\sigma_o}{r_{ij} - \sigma\left(\hat{\mathbf{u}}_i, \hat{\mathbf{u}}_j, \hat{\mathbf{r}}_{ij}\right) + \sigma_o}\right)^{12} - \left(\frac{\sigma_o}{r_{ij} - \sigma\left(\hat{\mathbf{u}}_i, \hat{\mathbf{u}}_j, \hat{\mathbf{r}}_{ij}\right) + \sigma_o}\right)^{6}\right] \quad (4.1)$$

The potential is essentially a modified version of the Lennard-Jones or 6-12 potential, which, in addition to the dependence on distance, r_{ij}, between the mesogens, also depends on their relative orientation described by vectors (directors) $\mathbf{u_i}$ and $\mathbf{u_j}$ (see inset of Figure 4.4(a)). With this approach one can successfully simulate the isotropic (Figure 4.4(b)), the nematic (Figure 4.4(c)), and the smectic (Figure 4.4(d)) phases. Timescales for coarse-grained simulations can be prolonged into the regime of microseconds.

At last, in the macroscopic regime simulations consider a continuum phase without any molecular detail, similar to classical fluid dynamics. Timescales lie in the range of milliseconds to seconds and sizes up to millimetres, thus describing the macroscopic behaviour of a liquid crystal sample, for example when applying an

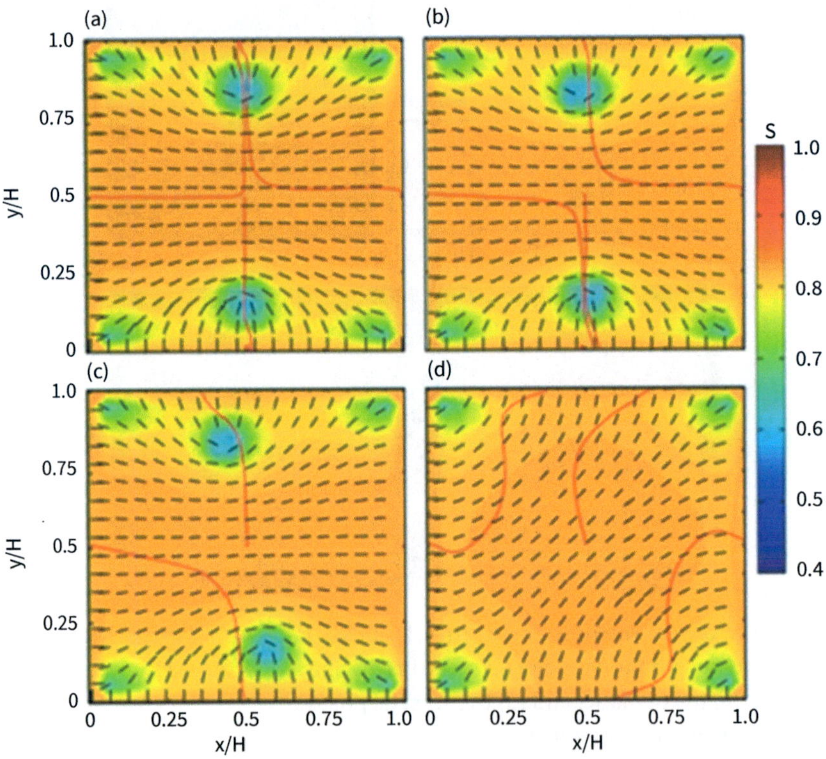

Figure 4.5 Continuum description of a nematic liquid crystal in a square cavity with metastable defects formed in the middle between the walls. The defects collapse and eventually yield four defects in the corners with zero topological charge overall. These types of simulations are carried out over spatial extensions in the millimetre range at timescales in the range of seconds.

(Reproduced with permission from ref. [8].)

electric or magnetic field to a liquid crystal in a sandwich cell, describing topological defects or director fields around micro-inclusions. The simulations of the director fields and reorientations are mainly done by a minimization of the free energy of the system. An illustrative example is shown in Figure 4.5 which shows a confined nematic phase in a cavity with a stationary lid at different times [8]. A metastable state with defects near the middle of the walls forms and collapses until a stable state is observed with defects in the four corners of the cavity and zero topological charge overall.

4.2 Hard rod simulations

The first simulations of lyotropic systems go back about 30 years and are based on Onsager's theory [9]. This was developed for hard discs and rods without any molecular interactions, being plainly based on steric interactions, i.e., excluded volume effects. Experimentally, such work goes back to the beginning of the twentieth century in studies related to plate- and rod-like inorganic and mineral liquid crystals. Here one should mention the investigations of Diesselhorst and Freundlich [10,11] on vanadium pentoxide, V_2O_5 (later revisited by Davidson et al. [12,13]) or the studies by Zocher [14] on AlO(OH). Other systems were later to follow, in particular clays and more recently synthesized colloidal rods and discs. These experimental systems and some possible applications will be covered in a later chapter of this text, while we here concentrate on computer simulations.

Figure 4.6 shows a simulated phase diagram of a hard rod system, including the biphasic regions [16]. Displayed are the stability regions of the isotropic, the nematic, the smectic, and the crystalline phases. The liquid crystalline phases are obtained once the aspect ratio of the rods reaches a certain value, indicating the necessity of shape-anisotropic colloidal particles to exhibit liquid crystalline behaviour, as expected.

An experimental example of a system simulated in Figure 4.6 would be goethite in water [17,18]. Goethite is a common mineral consisting of iron(III) oxide-hydroxide, α-FeO(OH), which has a velvety appearance and has been used since ancient times as a brownish pigment. The material forms nanorods of approximately 300 nm in length and 40 nm in width, with a relatively large polydispersity. Dispersed in water this would therefore be a colloidal lyotropic liquid crystal. A typical transmission electron microscopy (TEM) image of such goethite nanorods is shown in Figure 4.7(a) [19]. Some corresponding textures between crossed polarizers are depicted in Figure 4.7(b) as the concentration of goethite is increased from top to bottom. At low concentrations the isotropic phase (I) of largely diluted nanorods is observed, signified by the overall black appearance between crossed polarizers. For increasing concentration the transition to the nematic phase (N) is obtained, visible by the shown Schlierentexture. Further increase in concentration results in the formation of a smectic phase (Sm) before crystallization (C) is observed [18]. The described behaviour is evidenced by the corresponding X-ray diffraction

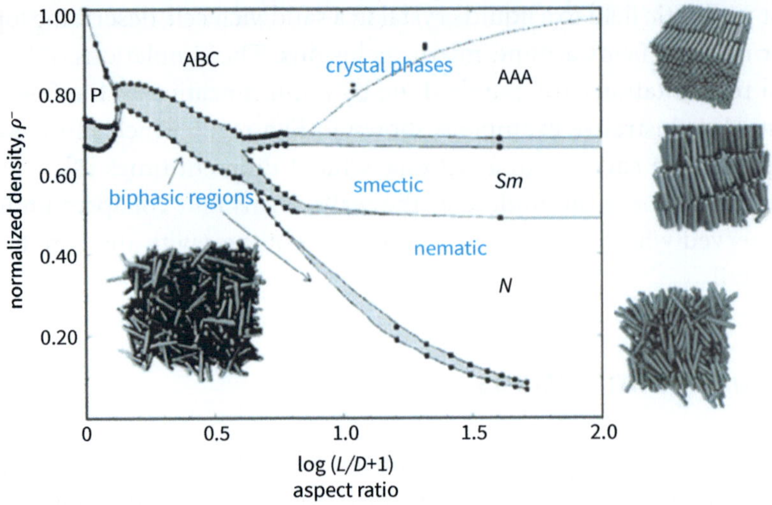

Figure 4.6 A simulated phase diagram of a lyotropic hard rod system yielding the isotropic, nematic, smectic, and crystalline phases. Note that at each transition a two-phase region is observed as expected from thermodynamic description. Such a simulation would for example qualitatively describe the behaviour observed for dispersed nanorods in water as the nanorod concentration and aspect ratio is changed.
(Reproduced with permission from ref. [15]).

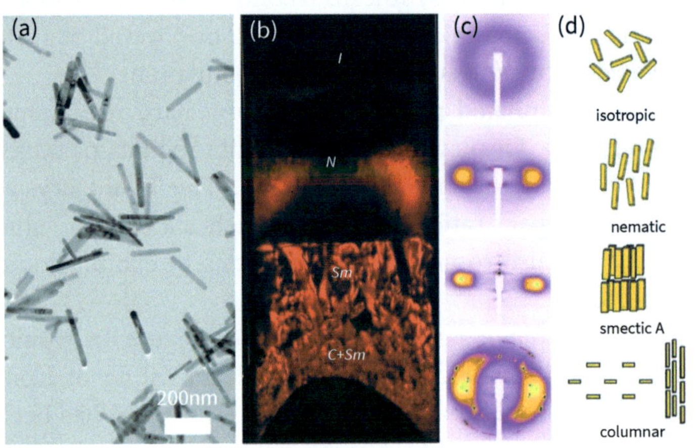

Figure 4.7 Example of an experimental lyotropic system which exhibits the phase diagram simulated in Figure 4.6 is goethite in water. (a) Shape anisotropic nanorods of goethite with an approximate typical length of 300 nm and width 40 nm. (Reproduced with permission from ref. [19].) (b) Liquid crystalline phase sequence for increasing goethite concentration in water from top to bottom. At low concentration the system is isotropic and thus appears black between two crossed polarizers. As the concentration increases the nematic phase is formed with birefringence appearing. Further increase in concentration leads to the smectic A phase and eventually crystallization. (c) The described phase behaviour is evidenced by X-ray diffraction. (d) Schematic illustration of the structural ordering of the goethite nanorods.
(Figure (d)reproduced with permission from ref. [15]).

diagrams of Figure 4.7(c) with the isotropic phase only showing liquid-like scattering as observed from the diffuse halo, the nematic phase exhibiting orientational ordering, to which positional ordering is added in the smectic phase [20]. Figure 4.7(d) depicts schematically the nanorod ordering discussed.

4.3 Gay–Berne simulations

Until about ten years ago the simulation of lyotropic liquid crystal phases from amphiphilic molecules via the formation of micelles was a true computational challenge, because the timescale at which this happens lies in the range of milliseconds,

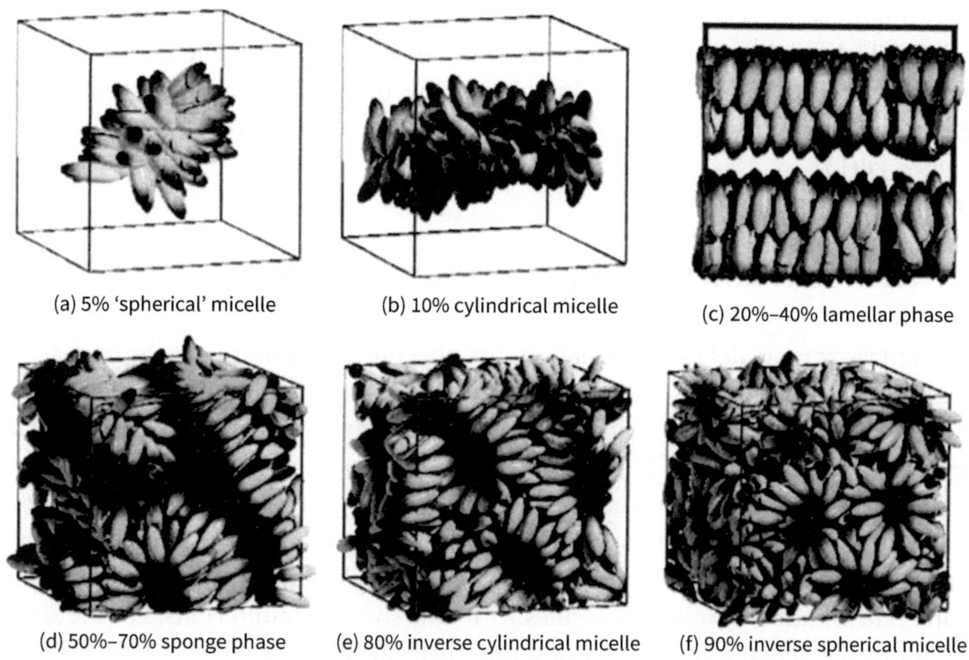

Figure 4.8 Structural features and phase transitions of a system of amphiphilic Gay–Berne particles for increasing concentration. (a) At low concentrations, but above the critical micelle concentration (CMC) the formation of isolated spherical micelles is observed. The system is in the isotropic phase with the micelles dispersed in water. (b) For increasing concentration of the amphiphilic ellipsoid the spherical micelles transform into cylindrical micelles. The system is still isotropic. (c) At a certain concentration and passing a two-phase region, the transition to the first lyotropic liquid crystalline phase is observed, in this case the lamellar phase. Between the formed bilayers the water is located. This structure exhibits self-organized, spontaneous order and is thus liquid crystalline and would appear bright between crossed polarizers due to its birefringence. (d) Next, the so-called sponge phase, a bilayer, bicontinuous structure, often of cubic symmetry, is observed, before (e) and (f): on further increase of the concentration inverse micelles are formed in the reverse order as they appear in the low-concentration regime.

(Reproduced with permission from ref. [21]).

too long for atomistic simulations. One of the earlier simulations of lyotropic behaviour was carried out by Michel and Cleaver [21], slowly increasing the concentration of a Gay–Berne particle. After crossing the critical micelle concentration (cmc) first sphere-like micelles are observed (Figure 4.8(a)), which, by increasing concentration, transition into a cylindrically shaped micelle (Figure 4.8(b). In this regime the hydrophilic head groups of an amphiphile or surfactant molecule would point towards the outside of the micelle to make contact with the water molecules, while the hydrophobic alkyl chains point towards the inside to avoid interactions with water. The transition to the first lyotropic liquid crystal phase is then obtained for even higher concentrations, in this case the lamellar phase (Figure 4.8(c)) and later a phase which is named the sponge phase (Figure 4.8(d)), a bicontinuous bilayer network structure, often of cubic symmetry. At very large concentrations one can then observe the reverse structures being formed where hydrophilic head-groups point to the inside of a reverse micelle, while the hydrophobic chains point to the outside. In this regime the cylindrical reverse micelles (Figure 4.8(e)) are formed at lower concentrations than the reverse spherical ones (Figure 4.8(f)).

4.4 Atomistic and coarse-grained simulations

In recent years more detailed simulations have become possible, due to the increased computing power at lower costs. Yet, fully atomistic simulations are still challenging and restricted in the number of molecules used. One example is shown in Figure 4.9 for a linear alkylbenzene sulfonate [22,23]. In part (a) one configuration of such a molecule is shown, where the sulfur atom is pictured in yellow and the oxygen atoms in red, forming the sulfonate head-group. A fully atomistic simulation is shown as a snapshot after 100 ns in Figure 4.9(b) where a single spherical micelle is clearly visible, surrounded by water molecules with blue sodium counterions. For the simulation of the actual lyotropic phases, one then has to resort to a coarse-grained model (Figure 4.9(c)), where the orange bead represents the sulfonate head-group connected to the yellow bead representing the phenyl ring. The chain of purple spheres illustrates the flexible alkyl chain. By increasing concentration the coarse-grained model first exhibits the hexagonal phase (top row of Figure 4.9(d)), which results from cylindrically shaped micelles, before at even higher concentrations the lamellar phase is formed (bottom row of Figure 4.9(d)).

The lamellar phase and the associated bilayer structures have been of especial interest in recent times, due to their relation to membranes and biological structures [24,25]. Molecular dynamics (MD) simulations of biological membranes are often carried out by first simulating the structure of the phospholipid bilayer where the composition is based on experimental work, for example the composition of a human cell membrane (Figure 4.10(a)). Besides following the equilibration during the simulation to thermodynamic equilibrium, one can also follow the dynamics

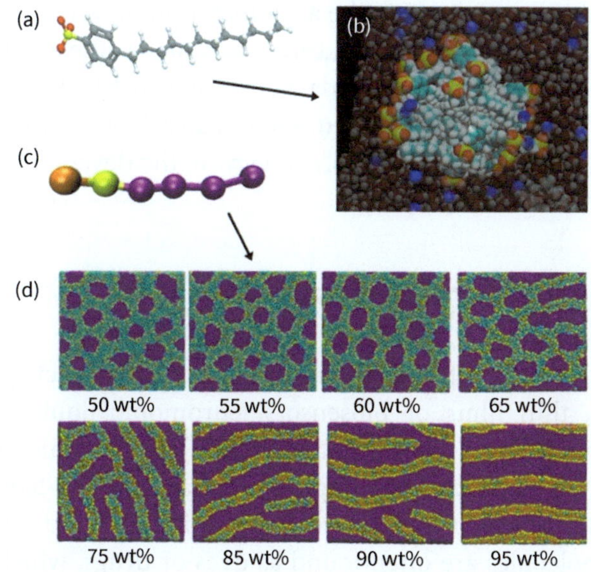

Figure 4.9 (a) Molecular structure of the alkylbenzene sulfonate investigated. (b) A snapshot of the full-atomistic simulation of a spherical micelle of this alkylbenzene sulfonate, surrounded by water molecules (grey and red) in the presence of sodium counterions (blue). (c) For a full simulation of the lyotropic phases one must resort to coarse-graining the molecule, with the sulfonate head-group (orange) connected to the phenyl ring (yellow) and the alkyl chain (purple). (d) The coarse-grained simulations then yield the hexagonal phase (top row) transforming into the lamellar phase (bottom row) for increasing molecule concentration.
(Reproduced with permission from refs. [22,23]).

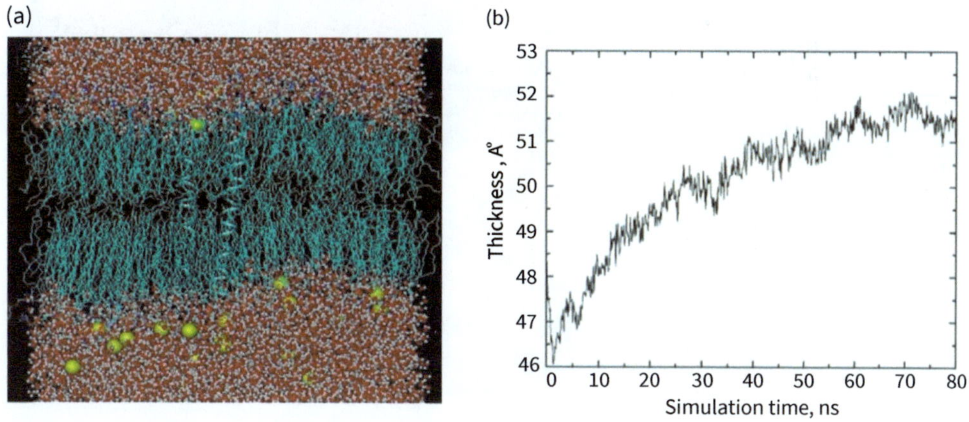

Figure 4.10 (a) Simulation of a phospholipid bilayer with the composition of a human cell membrane. (b) After thermodynamic equilibration in approximately 80 ns the thickness of the bilayer structure remains constant at about 5.2 nm which compares well to the 5.5 nm determined from experiments.
(Reproduced with permission from ref. [26]).

of the structure, due for example to the agglomeration of water molecules between the phospholipid bilayers, which provides a timescale for the equilibration process and a comparison with experimental data (Figure 4.10(b)). Further details of the membrane can by now also be included, such as a variable concentration of counterions, proteins, and protein channels. A review of the development of membrane simulations can be found in [27].

4.5 Simulation of chromonics

For other applications such as templates for nanoconstruction and synthesis of nanoparticles, thin films or biosensors, chromonic liquid crystals have in recent years moved into the centre of attention [28]. Chromonics are lyotropic liquid crystalline systems which are composed of plate- or disc-like molecules with hydrophilic groups attached to their sides which increase the solubility in water. These types of molecules are often found as dyes or drugs, which will also make the chromonic phases of interest for drug delivery and controlled drug release. A very common example is sunset yellow (SSY, molecular structure given in Figure 4.11(a)), which is found in many products as a food colouring and can be bought cheaply as a yellow powder (Figure 4.11(b)).

In terms of the thermodynamic behaviour, chromonics behave rather similar to conventional amphiphilic molecules. The formation of the mesogenic phase occurs for increasing concentration of the dye molecule and a relatively broad two-phase region has to be passed until the whole sample is liquid crystalline. In the case of

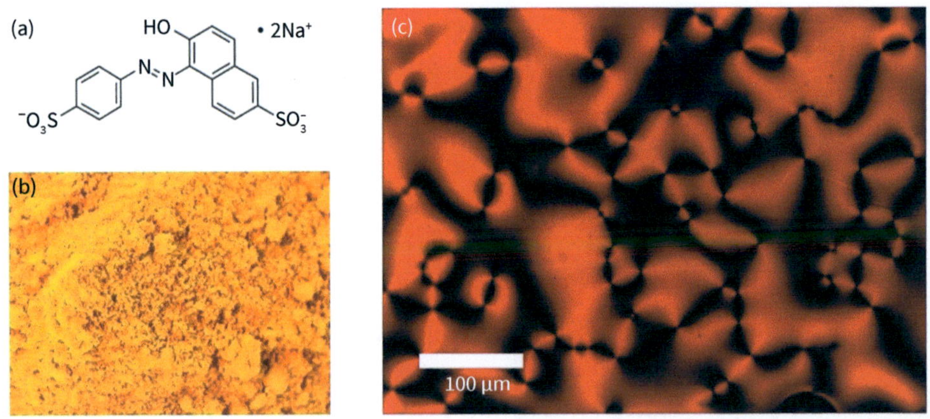

Figure 4.11 Chromonic lyotropic liquid crystals are often observed for dye or drug molecules. They possess a flat, plate-like molecular part and hydrophilic groups at their perimeter to aid solubility in water. (a) One of the standard chromonics is sunset yellow (SSY). (b) This is a dye molecule often used in food colouring. (c) At concentrations above 29 wt% in water SSY exhibits a fully developed nematic phase, here depicted with typical topological defects, called ±1/2 Schlieren-defects. Image of part (c) taken by Adam Draude.

sunset yellow the first phase to be observed is a nematic which transforms into a columnar phase at higher concentrations. At room temperature (20 °C) the isotropic phase of SSY in water exists for concentrations up to approximately 24% by weight, after which a two-phase region, I+N, is exhibited between 24 and 29 wt% before the whole field of view is nematic. Between crossed polarizers typical topological Schlieren-defects can be observed as shown in Figure 4.11(c). At high SSY concentrations of about 55 wt% a columnar phase is formed, again via a two-phase region, N-Col [29].

The general pathway how these plate-like molecules form liquid crystalline phases is via stack formation [30,31]. Molecules form columnar stacks with their hydrophilic peripheral groups ensuring good solubility with the water environment, as depicted in the top view of a stack in Figure 4.12(a). The stacks elongate and essentially mimic a more or less flexible rod surrounded by the solvent (Figure 4.12(b)). This process can be demonstrated by computer simulations, here with eight SSY molecules, where with increasing time in the ns regime the stacks grow from single molecules to dimers, trimers, and tetramers, which have formed after about 22 ns (Figure 4.12(c)). For large-scale simulations (or in real life) the length of the stacks would obviously be polydisperse to a certain degree. At low concentrations, like in

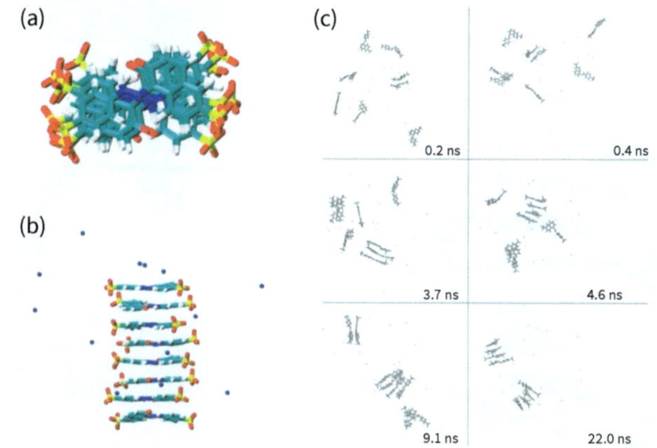

Figure 4.12 A snapshot of an atomistic simulation of SSY molecules forming a stacked aggregate (top view). (b) Side view of such a molecular stack, surrounded by water. At low concentrations these aggregates are randomly dispersed in the solvent. (c) A time-resolved series of snapshots to illustrate the aggregation from molecularly dispersed SSY molecules at very early times of 0.2 ns and the formation of dimers, trimers, and tetramers as time passes. At 22 ns the SSY molecules have associated into two tetramers. Continuing the simulation to times in excess of 200 ns will reveal the association of the two tetramers into a stack of eight molecules as shown in part (b). Note that all of these simulations describe the behaviour of small, independent aggregates forming in the isotropic phase. This simulation does not cover any lyotropic behaviour, but rather the mechanism on the molecular basis that will eventually lead to lyotropic liquid crystals.
(Reproduced with permission from ref. [30]).

Figure 4.12(c), the molecular stacks are oriented with their long axis in all directions with the same probability. Furthermore, SSY molecules can join and leave a given stack, similar to amphiphilic molecules joining and leaving a micelle. In addition, whole stacks can form or be destroyed at much longer relaxation times, and also two shorter stacks can combine to form one longer one. In the particular simulation discussed here, the two tetramer aggregates in Figure 4.12(c) did in fact merge to a stack of eight molecules after simulation times in excess of 200 ns.

This is the state where the isotropic phase is observed. Increasing the SSY concentration increases the stack length, but also the stack density, to a point where individual columnar stacks or 'rods' start interacting with each other. In the easiest picture one could describe this as simple steric interactions of the Onsager type, so that there will be a concentration where stacks will self-assemble, orient parallel to their neighbours, and start forming a nematic phase. Passing through the concentration regime of the two-phase region the nematic will grow at the expense of the isotropic phase until eventually only the nematic phase with its orientational order

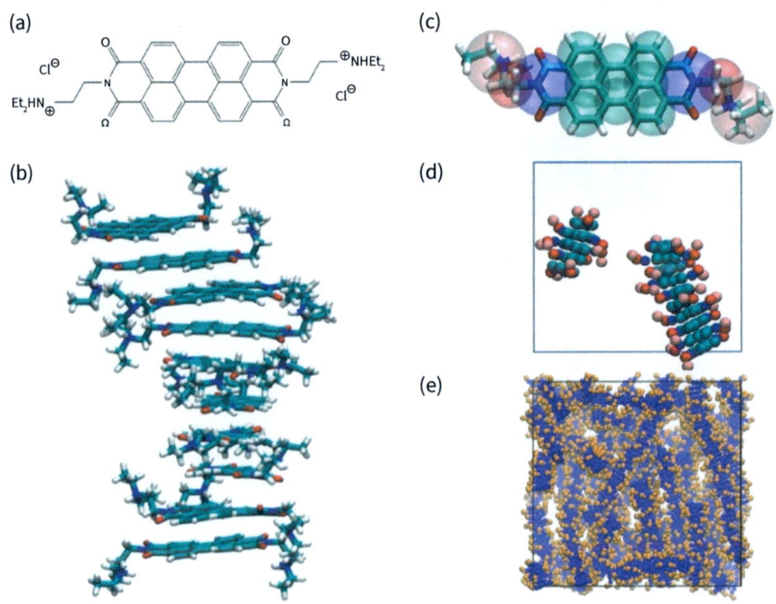

Figure 4.13 A hybrid approach to combine atomistic and coarse-grained models for an optimized description of chromatic lyotropic phases. (a) Molecular structure of the perylene dye, PER, under investigation. (b) Snapshot of a full atomistic simulation of a stack of eight PER molecules, which can be used to determine parameters to optimize the coarse-graining approach for further simulation of phase structures. (c) Schematic illustration of atomistic and coarse-grained molecular dye structure. (d) Coarse-grained simulation of the molecular stacking, which replicates the atomistic model closely. (e) Snapshot of the coarse-grained structure obtained for the chromonic nematic phase.
(Reproduced with permission from ref. [32]).

of stacks will be in existence. At this concentration, which is temperature dependent, the lyotropic nematic chromonic phase of sunset yellow has formed.

Very recently, Yu and Wilson [32] have carried out a large-scale study in terms of the number of molecules as well as time, where they compared a full-atom simulation of a perylene dye (PER) with novel approaches to match coarse-grained descriptions to the former results in order to capture both structure formation, self-organization, and the thermodynamics of a chromonic liquid crystal.

In Figure 4.13(a) the actual dye molecule under investigation is shown. These molecules can be simulated in an all-atom simulation where a snapshot of stacking eight such molecules is depicted in Figure 4.13(b). From these simulations a potential may be determined that describes the interactions between two adjacent molecules on an atomic level, which can then be used to develop an optimized coarse-grained description. The actual coarse-grained molecule model is schematically depicted in Figure 4.13(c), which is then used to simulate the stack formation, which corresponds very well to that of the all-atom calculation (Figure 4.13(d)). A long simulation run of the optimized coarse-grained model then yields a snapshot of the nematic phase as shown in Figure 4.13(e). It should be noted that using this hybrid approach of combining all-atom simulations with coarse-grained optimized procedures not only provides the nematic phase but at higher concentrations also the columnar phase of the chromonic LC in water.

4.6 Biaxial nematic liquid crystals

One of the timely aspects of current liquid crystal research is the evidencing and properties of the biaxial nematic phase. While this was first observed for lyotropic nematics more than four decades ago by Yu and Saupe [33], with other systems following [34,35] and including uniaxial to biaxial transitions [36], the occurrence of a biaxial nematic phase in thermotropic materials has been quite elusive and much discussed [37]. The difference between the uniaxial and the biaxial nematic phase is rooted in molecular shape anisotropy and the dynamics of the non-collective molecular rotations. This is schematically illustrated in Figure 4.14 where in the left part of the figure the uniaxial case and in the right part the biaxial case is shown.

The uniaxial nematic phase can be thought of as being composed of cigar-shaped molecules which fulfil a non-biased rotation around the long molecular axis. This leads to two observable refractive indices, $n_{\parallel}$ and $n_{\perp}$, with respect to a single optic axis. The biaxial nematic phase, on the other hand, is composed of brick-like molecules, which biases the molecular rotation, and three principal refractive indices are observed, $n_1 \neq n_2 \neq n_3$, leading to two optic axes.

In computer simulations of the thermotropic biaxial nematic phase and its phase diagram the shape anisotropy of the orthorhombic brick is taken as temperature independent. Figure 4.15 illustrates the results of such a simulation as a snapshot, depicted as a view from three orthogonal directions. This illustrates the biaxial

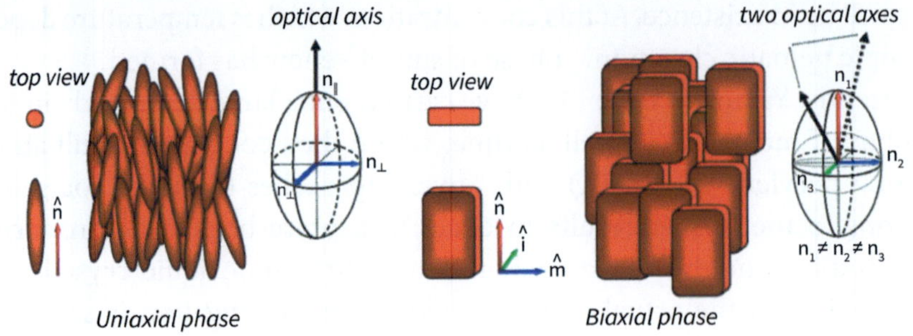

Figure 4.14 Schematic illustration of the uniaxial nematic (left) and the biaxial nematic phase (right).
(Reproduced with permission from ref. [38]).

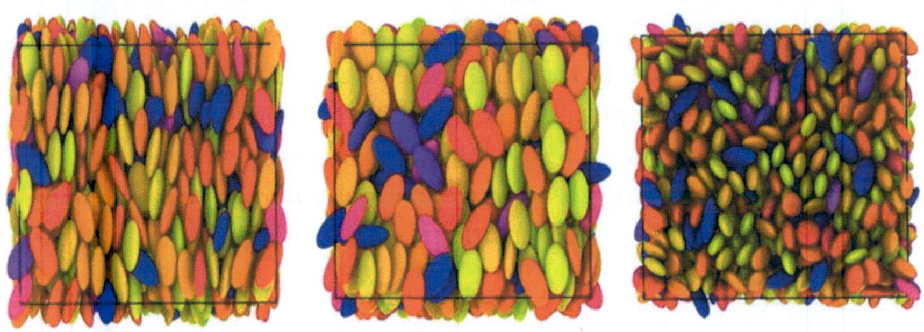

Figure 4.15 Illustrative results of a Gay–Berne simulation of a thermotropic biaxial nematic phase. The three parts of the image depict the obtained snapshot from three orthogonal directions.
(Reproduced with permission from ref. [40]).

nematic phase; nevertheless, the latter appears to be located in a rather small region of the phase diagram [39].

A more realistic view in comparison to experimentally observed behaviour may be the inclusion of a certain amount of polydispersity in length, σ_l, and interaction between neighbouring mesogens [41]. This leads, at least according to simulations, to a wider biaxial nematic phase, which forms at all times from a uniaxial nematic (see Figure 4.16).

We should note that to the present day there do not seem to be any convincing computer simulations of biaxial lyotropic behaviour that describe the phase diagram. It is suggested to modify the calculations to include the possibility of a variation of the shape anisotropy of the basic unit as a function of temperature and relative concentrations.

In conclusion, it can be said that since the first simulations of liquid crystals about three decades ago, massive improvements have been made, not only due to increased computing power and hardware, but also to continuous developments of software approaches and innovative mechanisms that allow for the simulation of a larger

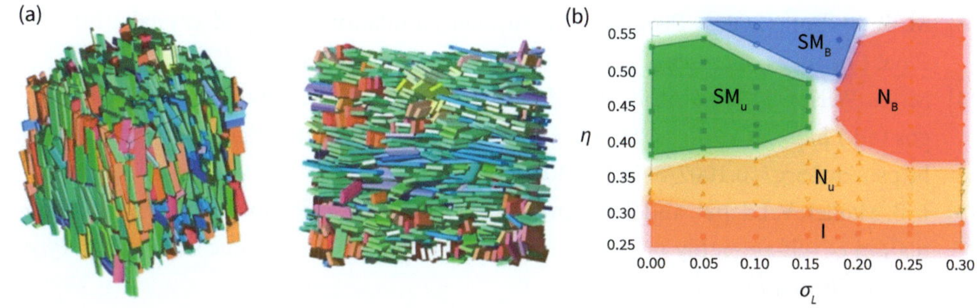

Figure 4.16 (a) Simulation snapshot of polydisperse brick-like mesogens forming a thermotropic biaxial nematic phase, shown from two different directions. (b) Obtained phase diagram as function of a length polydispersity parameter, σ_l.
(Reproduced with permission from ref. [41]).

number of molecules and longer timescales. While most of the simulations were initially done for thermotropic liquid crystals, lyotropics have increasingly featured in recent years, despite the problematic fact that molecules and molecular assemblies need to be simulated in an isotropic solvent, mainly water, which represents a considerable challenge. Of course, the simulation community is still quite far away from the holy grail of being able to simulate thermodynamic ensembles of molecules over timescales of micro- or even milliseconds. Nevertheless, much of the observed behaviour, even exotic phases like the recently discovered bent-core or twist-bend nematic phases, can nowadays be obtained by simulations with a reasonable number of coarse-grained molecules. The developments so far, together with an increased awareness of the biological functionality and applications of lyotropic LC systems, suggest that in the future a similar trend may be observed for lyotropic phases.

Finally, for readers who intend to delve deeper into the topic of computer simulations of liquid crystals in general, we can suggest several review articles which cover a range of different topics and provide references for further reading [2, 3, 6, 22, 42]. Furthermore, just very recently a first monograph on liquid crystal computer simulations by Zannoni has appeared [43], which provides the background and the details of different approaches as well as computing packages available. This book is not only suitable for beginning students in the field, but also for advanced researchers who use computer simulations in their work.

References

[1] M. Praprotnik, L. D. Site, and K. Kremer (2008). Multiscale simulation of soft matter: from scale bridging to adaptive resolution. *Annu. Rev. Phys. Chem.*, 59, 545–571.

[2] C. M. Care and D. J. Cleaver (2005). Computer simulation of liquid crystals. *Rep. Prog. Phys.*, 68, 2665–2700.

[3] M. R. Wilson (2005). Progress in computer simulations of liquid crystals. *Int. Rev. Phys. Chem.*, 24, 421–455.

[4] M. R. Wilson (1999). Atomistic Simulations of Liquid Crystals. In: D. M. P. Mingos (ed.) *Liquid Crystals I. Structure and Bonding*, vol 94, Springer: Berlin, pp 41–64. https://doi.org/10.1007/3-540-68305-4_2.

[5] J.-B. Maillet, V. Lachet, and P. V. Coveney (1999). Large scale molecular dynamics simulation of self-assembly processes in short and long chain cationic surfactants. *Phys. Chem. Chem. Phys.*, 1, 5277–5290.

[6] M. R. Wilson (2007). Molecular simulation of liquid crystals: Progress towards a better understanding of bulk structure and the prediction of material properties. *Chem. Soc. Rev.*, 36, 1881–1888.

[7] J. G. Gay and B. J. Berne (1981). Modification of the overlap potential to mimic a linear site–site potential. *J. Chem. Phys.*, 74, 3316–3319.

[8] J. P. Hernández-Ortiz, B. T. Gettelfinger, J. Moreno-Razo, and J. J. de Pablo (2011). Modeling flows of confined nematic liquid crystals. *J. Chem. Phys.*, 134, 134905.

[9] L. Onsager (1949). The effects of shape on the interaction of colloidal particles. *Ann. N. Y. Acad. Sci.*, 51, 627–659.

[10] H. Diesselhorst and H. Freundlich (1915). On the double refraction of vanadine pentoxydsol. *Phys. Z.*, 16, 419–425.

[11] H. Freundlich (1916). Die Doppelbrechung des Vanadinpentoxydsols. *Ber. Bunsenges. Phys. Chem.*, 22, 27–33.

[12] P. Davidson, A. Garreau, and J. Livage (1994). Nematic colloidal suspensions of V_2O_5 in water—Or Zocher phases revisited. *Liq. Cryst.*, 16, 905–910.

[13] O. Pelletier, P. Sotta, and P. Davidson (1999). Deuterium nuclear magnetic resonance study of the nematic phase of vanadium pentoxide aqueous suspensions. *J. Phys. Chem. B*, 103, 5427–5433.

[14] H. Zocher (1925). Über freiwillige Strukturbildung in Solen. (Eine neue Art anisotrop flüssiger Medien.). *Zeitschrift für Anorganische und Allgemeine Chemie*, 147, 91–110.

[15] H. N. W. Lekkerkerker and G. J. Vroege (2013). Liquid crystal phase transitions in suspensions of mineral colloids: new life from old roots. *Phil. Trans. R. Soc. A*, 371, 20120263.

[16] P. Bolhuis and D. Frenkel (1997). Tracing the phase boundaries of hard spherocylinders. *J. Chem. Phys.*, 106, 666.

[17] E. van den Pol, A. V. Petukhov, D. M. E. Thies-Weesie, D. V. Byelov, and G. J. Vroege (2009). Experimental realization of biaxial liquid crystal phases in colloidal dispersions of boardlike particles. *Phys. Rev. Lett.*, 103, 258301.

[18] E. van den Pol, D. M. E. Thies–Weesie, A. V. Petukhov, D. V. Byelov, and G. J. Vroege (2010). Uniaxial and biaxial liquid crystal phases in colloidal dispersions of board-like particles. *Liq. Cryst.*, 37, 641–651.

[19] S. Hinrichs, L. Grossmann, E. Clasen, H. Grotian genannt Klages, D. Skroblin, C. Gollwitzer, A. Meyer, and B. Hankiewicz (2020). Goethite nanorods: Synthesis and investigation of the size effect on their orientation within a magnetic field by SAXS. *Nanomaterials*, 10, 2526.

[20] E. van den Pol, A. V. Petukhov, D. M. E. Thies-Weesie, D. V. Byelov, and G. J. Vroege (2010). Liquid crystal phase behavior of sterically-stabilized goethite. *J. Coll. Interf. Sci.*, 352, 354–358.

[21] D. J. Michel and D. J. Cleaver (2007). Coarse-grained simulation of amphiphilic self-assembly. *J. Chem. Phys.*, 126, 034506.

[22] M. R. Wilson, G. Yu, T. D. Potter, M. Walker, S. J. Gray, J. Li, and N. J. Boyd (2022). Molecular simulation approaches to the study of thermotropic and lyotropic liquid crystals. *Crystals*, 12, 685.

[23] S. J. Gray, M. Walker, R. Hendrikse, and M. R. Wilson (2023). Investigating anionic surfactant phase diagrams using dissipative particle dynamics: Development of a transferable model. *Soft Matter*, 19, 3092.

[24] S. J. Marrink, A. H. de Vries, and A. E. Mark (2004). Coarse grained model for semiquantitative lipid simulations. *J. Phys. Chem. B*, 108, 750–760.

[25] S. Dey and J. Saha (2017). Solvent-free, molecular-level modeling of self-assembling amphiphiles in water. *Phys. Rev. E*, 95, 023315.

[26] A. A. Shahinyan, P. K. Hakobyan, L. H. Arsenyan, and A. H. Poghosyan (2012). The study of lyotropic liquid crystal structure using the molecular dynamics simulation method. *Mol. Cryst. Liq. Cryst.*, 561, 155–169.

[27] S. J. Marrink, V. Corradi, P. C. T. Souza, H. I. Ingólfsson, D. P. Tieleman, and M. S. P. Sansom (2019). Computational modeling of realistic cell membranes. *Chem. Rev.*, 119, 6184–6226.

[28] R. Bosire, D. Ndaya, and R. M. Kasi (2021). Recent progress in functional materials from lyotropic chromonic liquid crystals. *Polym. Int.*, 70, 938–943.

[29] H.-S. Park, S.-W. Kang, L. Tortora, Y. Nastishin, D. Finotello, S. Kumar, and O. D. Lavrentovich (2008). Self-assembly of lyotropic chromonic liquid crystal sunset yellow and effects of ionic additives. *J. Phys. Chem. B*, 112, 16307–16319.

[30] F. Chami and M. R. Wilson (2010). Molecular order in a chromonic liquid crystal: a molecular simulation study of the anionic azo dye sunset yellow. *J. Am. Chem. Soc.*, 132, 7794–7802.

[31] D. J. Edwards, J. W. Jones, O. Lozman, A. P. Ormerod, M. Sintyureva, and G. J. T. Tiddy (2008). Chromonic liquid crystal formation by edicol sunset yellow. *J. Phys. Chem. B*, 112, 14628–14636.

[32] G. Yu and M. R. Wilson (2022). Molecular simulation studies of self-assembly for a chromonic perylene dye: All-atom studies and new approaches to coarse-graining. *J. Mol. Liq.*, 345, 118210.

[33] L. J. Yu and A. Saupe (1980). Observation of a biaxial nematic phase in potassium laurate-1–decanol-water mixtures. *Phys. Rev. Lett.*, 45, 1000.

[34] R. Bartolino, T. Chiaranza, M. Meuti, and R. Compagnoni (1982). Uniaxial and biaxial lyotropic nematic liquid crystals. *Phys. Rev. A*, 26, 1116.

[35] E. A. Oliveira, L. Liebert, and A. M. F. Neto (1989). A new soap/detergent/water lyotropic liquid crystal with a biaxial nematic phase, *Liq. Cryst.*, 5, 1669–1675.

[36] A. Saupe, S. Y. Xu, S. Plumley, Y. K. Zhu, and P. Photinos (1991). Micellar structures in lyotropic liquid crystals and phase transitions. *Physica A*, 174, 195–207.

[37] G. R. Luckhurst (2001). Biaxial nematic liquid crystals: Fact or fiction? *Thin Solid Films*, 393, 40–52.

[38] M. Lehmann(2009). The dawn in the area of thermotropic biaxial nematics. Symposium on Biaxial Liquid Crystals, Kent, 27–28 October 2008. *Liquid Crystals Today*, 18, 31–34.

[39] R. Berardi, L. Muccioli, S. Orlandi, M. Ricci, and C. Zannoni (2008). Computer simulations of biaxial nematics. *J. Phys.: Condens. Matter*, 20, 463101.

[40] S. Orlandi, L. Muccioli, and R. Berardi. From rod-like to disc-like Gay–Berne biaxial nematics and back. *Liq. Cryst.*, 45, 2400–2415.

[41] E. M. Rafael, D. Corbett, A. Cuetos, and A. Patti (2020). Self-assembly of freely-rotating polydisperse cuboids: Unveiling the boundaries of the biaxial nematic phase. *Soft Matter*, 16, 5565.

[42] M. P. Allen (2019). Molecular simulation of liquid crystals. *Mol. Phys.*, 117, 2391–2417.

[43] C. Zannoni (2022). *Liquid Crystals and their Computer Simulations.* Cambridge: Cambridge University Press.

5
Amphiphilics

5.1 Introduction

Lyotropic liquid crystals (LLC) may be formed mixing amphiphilic molecules and a solvent. The stabilization of this state of matter depends on the relative concentrations of the different components of the mixture and the temperature. The electrostatic interaction between molecules, or even parts of the molecules, is the driving force responsible for the stabilization of a particular molecular structure. To form a liquid crystalline structure, molecular segregation occurs on the nano-scale, mainly due to the electric affinity between molecules of the solvent and specific parts of the amphiphiles.

In this chapter we discuss the main types of amphiphilic molecules present in LLC, the structures identified, and typical phase diagrams. In particular nematic phases are described since it is only in LLC that the three nematic phases, two uniaxial and one biaxial, have been observed.

5.2 Types of amphiphilic molecules

The main characteristic of an amphiphilic molecule is the presence of two parts with antagonistic electrostatic affinity. One of these parts shows affinity with polar solvents and the other with nonpolar solvents. These interactions are commonly named *hydrophobic* and *hydrophilic*, evidencing the fact that electrically similar molecules (or part of a molecule) tend to segregate themselves from other molecules (or parts of a molecule) to be together.

Before we present the different types of amphiphiles, let us discuss an illustrative everyday example that explains this effect. When we have oil on our hand and we want to remove it, just water does not help us in this task. Oils are mainly hydrocarbons (nonpolar molecules) and water is a polar solvent. So, mixing oil and water on our hands will only spread the oil more and the hand will continue to stay dirty. However, if we take a soap, that is made of amphiphilic molecules, when we mix the oil, water, and the soap in our hands, superstructures of amphiphilic molecules, with the polar part of the molecules in the outer region, will be formed. The oil will be placed inside these superstructures and water will wash out the oil and the soap from our hands. This nano-segregation process is also the basis of the formation of the lyotropic liquid crystalline structures and phases.

Lyotropic Liquid Crystals. Ingo Dierking and Antônio Martins Figueiredo Neto, Oxford University Press.
 DOI: 10.1093/9780191924927.003.0005

Different types of amphiphiles exist, natural or chemically synthesized [1], and some examples are given in Figure 5.1: anionic (a), cationic (b), zwitterionic (c), detergents (d), anelydes [2] (e), gemini surfactants [3] (f), rigid spiro-tensiles [4] (g), phospholipids (h), facial amphiphiles [5] (i), and bolaamphiphiles [6] (j).

Amphiphilic molecules are employed in many areas [7], like in food products, cleaning detergents, cosmetic and pharmaceutical industries, and mineral and petroleum processing. A striking action of surfactants is their ability to strongly modify the surface properties of interfaces. At the water–air or oil–water interface, they reduce the surface tension, changing the wettability of a surface. This property allows many technological applications and has also opened interesting areas of basic scientific research.

Another type of amphiphiles are the block polymers (Figure 5.1g), macromolecules of high molecular weight, with different blocks that present hydrophobic or hydrophilic behaviour [8–10]. Nano-segregation occurs in the presence of solvents. The variety of molecules synthesized is high and depends on the interest in research and technological applications. An advantage of using this type of amphiphile is its high chemical stability [11].

The use of amphiphilic molecules in different applications seems to be far from an end. Recently, amphiphilic methacrylate copolymers with polysiloxane and mixed poly(oxyethylene)—perfluorohexyl side chains were synthesized [12]. This material, as part of a nanocomposite, was shown to present long-term marine antifouling performance. Libraries of organic amphiphiles were constructed, enlarging the frontiers of these molecules [13]. Other applications reported for these types of molecules are as anti-bacterial and anti-fungal agents and anti-corrosion on mild steel [14].

5.3 Self-assembled systems

The nanoscale segregation of amphiphilic molecules (AM) in solvents occurs, mainly, due to the electrostatic interaction between the different parts of the amphiphilic molecules and the solvent. When AM are mixed with a solvent, at small concentrations, monomers exist in the solvent but, after crossing a concentration threshold named the critical micellar concentration (CMC), the molecular segregation takes place and AM aggregates start to be formed, still in the presence of isolated AM [15,16]. Even after the CMC has been crossed, isolated AM are present in the mixture besides the micelles. The typical exchange time for an isolated AM and the micelle-incorporated situation is about 10^{-5}–10^{-3} s [17]. After the CMC, the number of micelles increases with respect to that of the AM. The CMC depends on the temperature of the mixture and the addition of substances that affect the interaction of the solvent molecule and the AM, e.g., salts [18]. The CMC dependence on temperature originated the concept of critical micellar temperature (CMT).

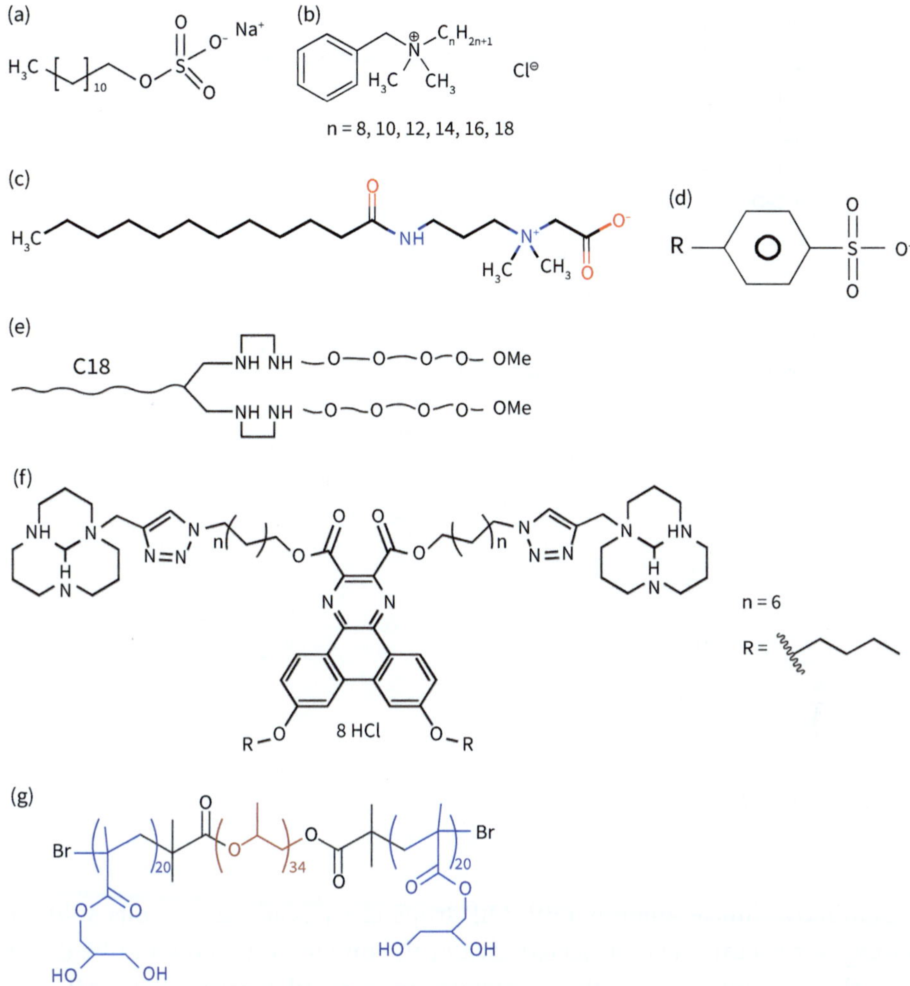

Figure 5.1 Examples of amphiphilic molecules (a) sodium dodecyl sulfate (SDS), anionic; (b) benzalkonium chloride, cationic; (c) lauramidopropyl betaine, zwitterionic; (d) detergent, sodium alkylbenzene sulfonate; (e) annelides, ME is a metallic ion; (f) DEDPP-4. 5,6-diphenylpyrazine-2,3-diester, decorated with triazole-[12]aneN_3, gemini; (g) spiro tensile—a triblock copolymer of propylene oxide (central block) and glycerolmethacrylate (terminal blocks); (h) phosphatidylcholine; (i) facial—functionalized 4,4″,2′-trialkoxyterphenyl; (j) 12-cyclohexyl-DQA:10-10′-(dodecane-1,12-diyl)bis(9-amino-1,2,3,4-tetrahydroacridinium) dichloride, bolaamphiphile.

Different experimental methods are used to determine the CMC and what was found is that each method furnishes a particular value for this parameter, leading researchers to identify a concentration interval to define the CMC [19]. Among these methods we can point out the measurements of the osmotic pressure, turbidity, surface tension, molar conductivity, and voltammetry. Recently, a new method to determine the CMC was proposed by using video microscopy

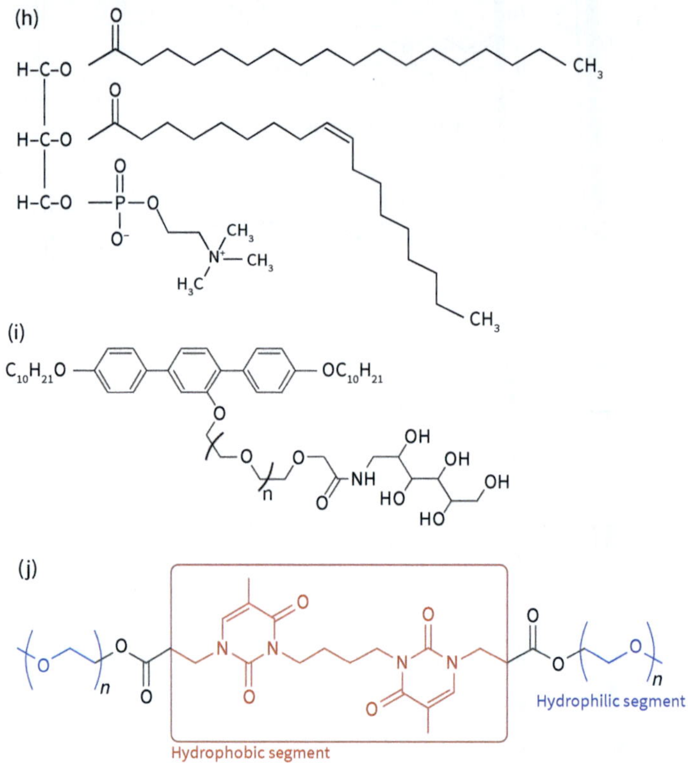

Figure 5.1 *Continued.*

and quantitative image analysis [20]. Oil drops are spread on aqueous solutions containing surfactants. The projected area of the oil lens was shown to be dependent on the concentration of the surfactant, in particular across the CMC. The method consists of plotting the maximum projected area occupied by oil drops spreading on aqueous surfactant solutions as a function of the surfactant concentration. Across the CMC this area shows an increase and the CMC can be measured (Figure 5.2).

In self-assembled systems, molecules organize themselves through noncovalent interactions (e.g., dipole–dipole interactions, Van der Waals forces, and hydrogen bonds).

Depending on the relative concentrations of the different components of the lyotropic mixture, direct or inverted molecular aggregates may be formed. Another relevant parameter affecting the aggregate polymorphism are the volumes occupied by the hydrophobic and hydrophilic parts of the amphiphile [21]. Figure 5.3 shows sketches of a typical amphiphilic molecule (a) and the available volumes for the hydrophobic and hydrophilic parts of it. The interfacial curvature of the molecular aggregate is strongly affected by the aspect of these volumes: (b)—e.g., lysophosphatidylcholine and (c) direct micelles and bicontinuous structures are favoured; (d)—e.g., phosphatidylcholine lamellar structures are favoured; (e)—e.g., glycerol monooleate inverted micellar and bicontinuous structures are favoured. Figure 5.4

shows representations of direct (a) and inverted spherical micelles (b). In both cases the hydrophilic part of the amphiphilic molecule is in contact with the polar solvent (e.g., water), and the hydrophobic part is in contact with the nonpolar solvent (e.g., oil).

The ability of biocompatible amphiphilic molecules to self-assemble in stable nano- and meso-structures allowed their use in drug delivery for actual applications in therapeutics and diagnostics [21]. Since the lyotropic structure has regions with different electric properties, target molecules can be dispersed in these regions

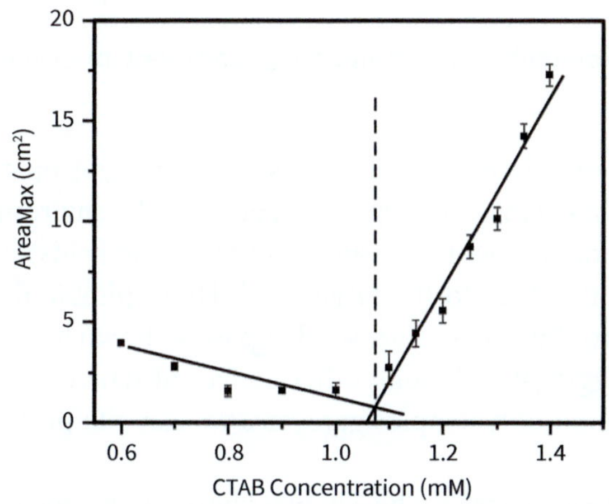

Figure 5.2 Maximum projected area occupied by oil drops spreading on aqueous CTAB solutions as a function of the CTAB concentrations. The intersection of the straight lines defines the CMC.

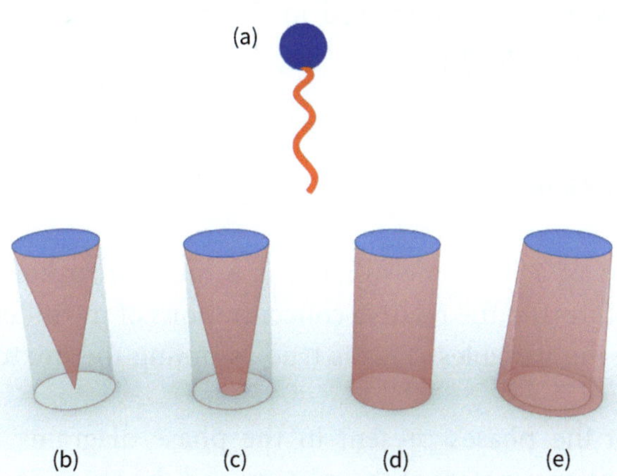

Figure 5.3 Sketches of a typical amphiphilic molecule (a) and the available volumes for the hydrophobic (red) and hydrophilic (blue) parts of it; (b) and (c) direct micelles and bicontinuous structures are favoured; (d) lamellar structures are favoured; (e) inverted micellar and bicontinuous structures are favoured.

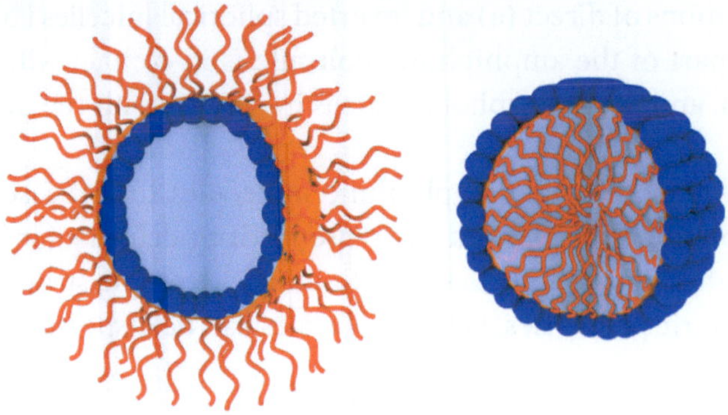

Figure 5.4 Representation of an inverted (a) and direct (b) spherical micelles.

and transported to the place in the living tissue by the lyotropic molecular aggregate. Since the stability of these aggregates depends on the environment and external actions (e.g., temperature, light, pressure, and magnetic fields), they can be destabilized and deliver the drug to the target [22]. Hydrophobic (e.g., paclitaxel and diazepam) and hydrophilic (e.g., glucose) drugs were shown to be successfully delivered by lyotropic aggregates. Among other active substances usually employed in this type of application of lyotropic aggregates we can cite proteins, peptides, and nucleic acids.

As will be discussed later in the chapter, besides the formation of micelles, amphiphilic molecules in a solvent may form complex bicontinuous structures, like the sponge (or L_3) phases [23]. This phase is optically isotropic, structured in a bicontinuous bilayer arrangement. This bilayer structure divides space everywhere into an inside and an outside region. An inverted sponge phase was also experimentally observed in the mixture sodium dodecyl sulfate/1-pentanol/cyclohexane/water [24].

5.4 Phase diagrams

The phase diagram of a lyotropic mixture is a multidimensional representation where the temperature and the relative concentrations of the different components of the mixture are the variables. This is true assuming the mixture is at constant pressure.

The topology of the phases present in the phase diagrams is very rich and depends on different parameters of the mixture and the characteristics of the main amphiphile. Among these characteristics we may cite the relation between the volumes available for the amphiphile polar head and that for the hydrophobic chain. If the polar head occupies a larger volume with respect to that of the hydrophobic chain, molecular aggregates with higher surface curvatures are stabilized, favouring the appearance of micellar isotropic and/or cylindrical aggregates. This tendency may be expressed by the parameter CPP (critical packing parameter), defined as

CPP = V/Ad, where V, A, and d are the hydrophobic chain volume of the amphiphile, the cross-sectional area of the hydrophilic head group, and the hydrophobic chain length, respectively [16]. Roughly, when CPP < 1, direct structures (e.g., micellar, cubic, sponge, and hexagonal) are formed. Inverted structures are favoured when CPP > 1. The lamellar phase is favoured when CPP = 1. This criterion has to be taken as a rough indication of the expected structure since lyotropics are very sensitive to the addition of new components to the mixture, e.g., electrolytes and co-surfactants.

For practical reasons, different 'cuts' of the multidimensional phase diagram, named partial phase diagrams, are commonly plotted. These cuts are particular surfaces of the multidimensional phase diagram of the mixture. Some examples are given in the following.

5.4.1 Partial isothermal phase diagram

Figure 5.5 shows a triangular representation of the isothermal phase diagram of the ternary mixture of sodium octanoate/octanoic acid/water [25,26]. The different phases are represented in regions delimited by lines. It is important to notice that when the transitions between phases are of first order, there is not a single line between the phases, but regions where both phases coexist. This type of representation is particularly useful when experimentalists are interested in the phase behaviour of the mixture at room temperature. Triangular representations were widely employed in the soap and cosmetics industry, with the different phases roughly identified by their textures in the polarized-light microscope. When the lyotropic mixture has more than three components this representation is not possible, except when one of the components of the mixture has a fixed concentration.

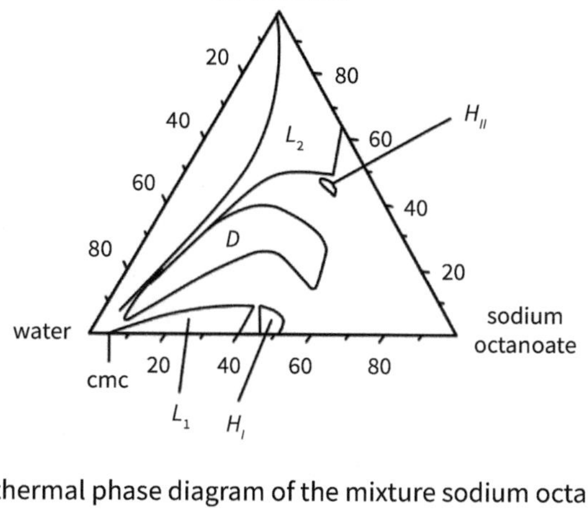

Figure 5.5 Partial isothermal phase diagram of the mixture sodium octanoate/octanoic acid/water, adapted from [25,26]. The labels L_1, L_2, D, H_I, and H_{II} represent the direct, inverted micellar phases, the lamellar phase, and the direct and inverted hexagonal phases, respectively.

5.4.2 Partial two-dimensional phase diagram

Figure 5.6 shows a sketch of a partial two-dimensional (2D) phase of a binary mixture amphiphile/water, with the typical phases encountered as a function of the temperature and relative concentration of the amphiphilic molecule [27]. The Krafft temperature line at low temperatures separates the liquid crystalline regions from the hydrated crystalline domain. In the left side of the phase diagram, at small amphiphile concentration, isolated molecules in water are encountered. Increasing the amphiphile concentration, the mixture reaches the CMC and the self-aggregation process occurs, giving rise to the isotropic micellar solution. By increasing the amphiphile concentration further, ordered mesophases are stabilized (e.g., nematic, hexagonal, and lamellar). In the right side of the phase diagram, for higher concentrations of the amphiphile, inverted structures may be stabilized.

Another usual way to represent 2D phase diagrams of lyotropic mixtures with more than two components is to fix the concentration (e.g., in Mol%) of all the components of the mixture, except two. An example is depicted in Figure 5.7, where the isotropic and the three nematic phases are present. The mixture is composed of potassium laurate/1-decanol (at a fixed concentration)/D_2O [28]. This representation is particularly useful to study the transitions between the different phases present in the multidimensional phase diagram of the mixture. As the temperature of the sample is a parameter that experimentalists may control with high precision (e.g., 0.001 °C in current laboratories), the characteristic of the phase transition from the point of view of statistical mechanics can be studied [1].

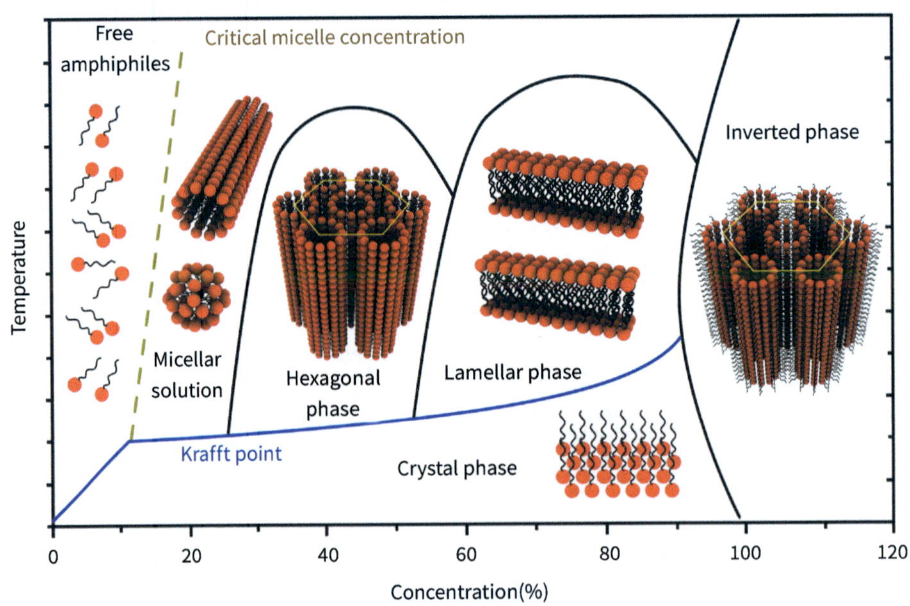

Figure 5.6 Sketch of a partial 2D phase diagram of a mixture amphiphile/water. (Reproduced with permission from ref. [27]).

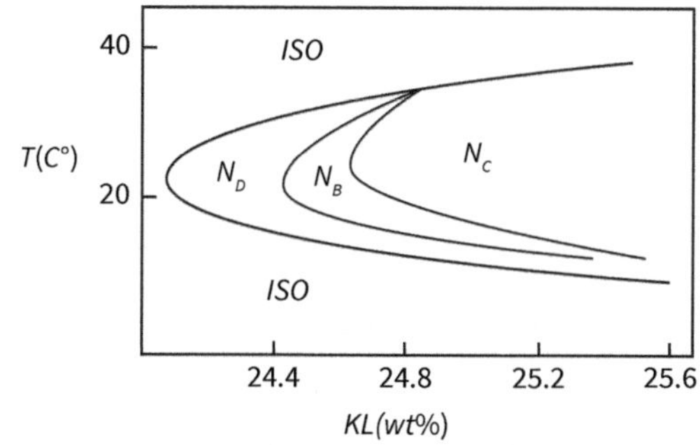

Figure 5.7 Sketch of the 2D partial phase diagram of the mixture potassium laurate (KL)/1-decanol(fixed at 6.24 wt%)/D_2O. ISO, N_D, N_B, and N_C stand for the isotropic, discotic nematic, biaxial nematic, and calamitic nematic, respectively.

5.5 Nematic mesophases

At the beginning of the 1970s, two seminal papers were published by Freiser [29] and Alben [30] predicting, from the theoretical point of view, based on symmetry arguments, the existence of three nematic phases. In particular, Alben calculated the phase diagram of a fluid of biaxial particles, revealing the characteristics of the transitions between these phases. Using the mean-field Landau-type approach, the transitions between the nematic phases were shown to be of second order, and the one from the nematic to the isotropic phase of first order. A Landau point, where two second-order transition lines encounter a first-order line, was also predicted [31].

Despite the fact that uniaxial nematic phases were found in lyotropic mixtures in the 1960s [32], it was only in 1980 that the biaxial nematic phase was encountered in a ternary lyotropic mixture [28]. After that, experimentalists investigated other lyotropic mixtures and it was possible to establish some experimental conditions for the mixture to stabilize the N_B phase [33]: i) at least two types of amphiphilic molecules must be present, with alkyl chain lengths differing by about two CH_2; ii) considering the interactions between head-groups of the surfactants and the counterions/ions of strong electrolytes (in terms of their kosmotropic and chaotropic characteristics), we have to choose the surfactant and the electrolyte with slightly opposite characteristics.

To the present day there is no clear evidence of a biaxial nematic phase in thermotropic liquid crystals. This is an intriguing fact; however, some aspects characteristic of lyotropics could bring some hints to understanding it. A striking characteristic of lyotropics is the ability to 'tune' the micelle shape anisotropy as a function of the temperature and the relative concentrations of the mixture [1,33]. This shape anisotropy is also dependent on the nano-segregation of the different amphiphilic molecules present in the micelle. This segregation was evidenced by a contrast neutron scattering experiment [34] in the mixture of potassium

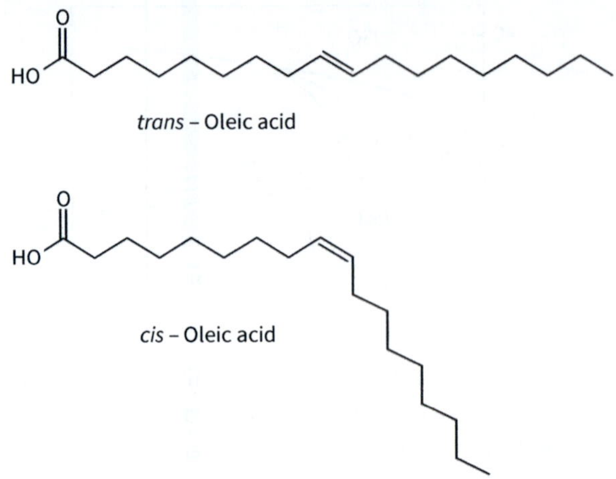

Figure 5.8 Oleic acid in cis and trans configurations.

laurate/decanol/water, indicating the presence of intrinsically biaxial micelles in a uniaxial calamitic nematic phase. Assuming a model of intrinsically biaxial micelles (IBM) in the nematic phases (this model will be discussed later), the micelles could be pictured as a flattened prolate ellipsoid, with main-axes dimensions A′, B′, and C′ [1]. X-ray scattering experiments show that the shape anisotropy in the biaxial nematic phase (mixture of potassium laurate/decanol/water) is about A′: B′: C′=3:2:1 [35]. Moreover, this shape anisotropy continuously varies as a function of the temperature and, depending on the relative concentrations of the amphiphiles (surfactant and co-surfactant), reaches the one that stabilizes the biaxial nematic phase. This fact was verified also in other lyotropic mixtures presenting the three nematic phases [33].

Another aspect that supports the idea of modification of the micellar shape anisotropy in lyotropics is the presence of an isotropic phase below the temperature of the uniaxial nematic phase [28]. Just taking into account classical thermodynamics, a system constituted of objects of fixed shape and shape anisotropy (e.g., prolate ellipsoidal micelles), the ordered phase (in this case the uniaxial nematic) is expected to appear at temperatures smaller with respect to that of the disordered (isotropic) phase. The isotropic phase encountered in many lyotropic phase diagrams at lower temperatures with respect to the ordered phases can be understood considering *l'effet cauchout* (i.e., the rubber effect). At higher temperatures, the carbonic chains present a higher propensity to show cis configurations, squeezing the micelle bilayer dimension (smaller values of C′ in the framework of the IBM model; see Figure 5.8). On the other hand, at lower temperatures, the trans configuration is favoured, which increases the bilayer thickness and, consequently, decreases the micellar shape anisotropy. This smaller shape anisotropy leads to the stabilization of the low-temperature isotropic phase.

The phase diagram of mixtures with one amphiphile (e.g., potassium laurate/KCl/water) in the nematic region shows only one uniaxial nematic phase. For

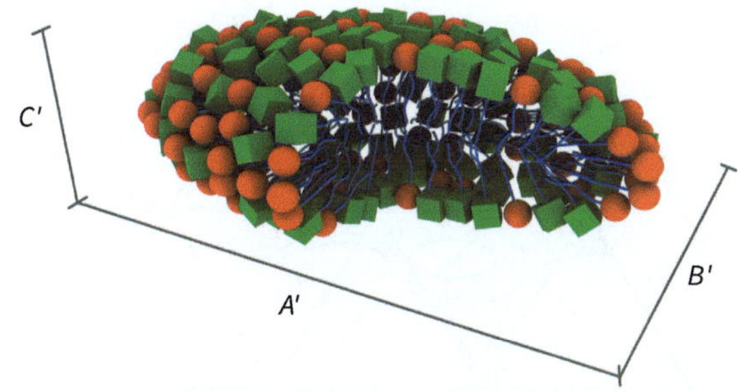

Figure 5.9 Sketch of an orthorhombic micelle (flattened prolate ellipsoid) with axis dimensions A′, B′ and C′. The main amphiphilic bilayer is represented by C′. The red spheres and the green cubes represent the main amphiphile and the co-surfactant, respectively. The carbonic chains are represented in blue.

symmetry reasons we expect that the micelles in these mixtures will have a uniaxial shape, prolate or oblate ellipsoids in the N_C and N_D nematic phases, respectively. As stated before, there is no evidence of mixtures with only one amphiphile showing the N_B phase.

The IBM model is based on both theoretical and experimental arguments. From the theoretical point of view, the seminal paper from Alben evokes the possibility of having both uniaxial and the biaxial nematic phase in a fluid of biaxial particles [30]. This simple and elegant idea is inherent of the Landau theory of phase transitions assuming that the order parameter that characterizes the phases is a second-rank tensor and a rotationally invariant Hamiltonian. Transitions between nematic phases are of second order and from nematic to isotropic of first order. From the experimental side, X-ray diffraction [35] and neutron scattering [34] experiments support the idea of intrinsically biaxial micelles (of orthorhombic symmetry) in the three nematic phases. The contrast neutron experiments reveal that the main amphiphile concentrates more in the curved surfaces of the micelles and the co-surfactant molecules are preferentially located in the flat surfaces of the micelles (Figure 5.9). According to the IBM model the different phases are originated by different orientational fluctuations of the micelles along the three orthogonal two-fold symmetry axes attached to the micelles. Orientational fluctuations that degenerate the axis parallel to the dimension A′ (C′) originate the N_C (N_D) phase. Small amplitude orientational fluctuations along the three orthogonal axes originate the N_B phase. Measurements of the optical birefringences show that the nematic-to-nematic phase transitions are of second order [36], in agreement with Alben's prediction. Details about the measurements of the optical birefringences of nematic lyotropic liquid crystals are given in Chapter 3 (laser conoscopy).

The elasticity of lyotropic nematics was investigated in the 1980s by Haven and co-workers in a mixture of decylammonium chloride/ammonium chloride/water [37]. Interestingly, the splay (k_{11}) and bend (k_{33}) elastic constants of lyotropic nematics measured were of the same order of magnitude of those from the thermotropic nematic MBBA (methoxybenzylidene butylaniline), about 10 pN. On the other

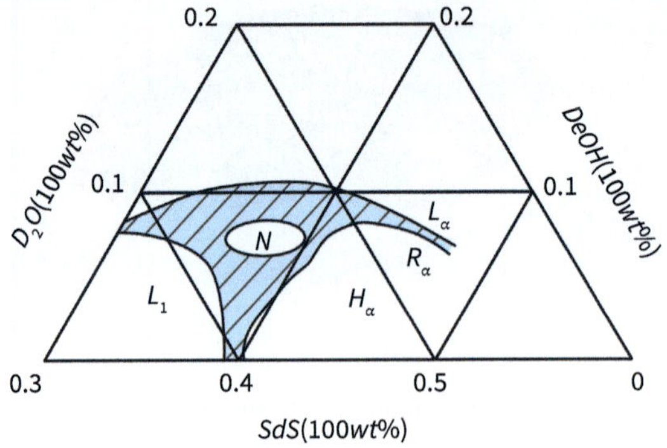

Figure 5.10 Triangular phase diagram of the mixture sodium decylsulfate(SdS)/decanol(DeOH)/D_2O, with the locus of the nematic phases. The labels L_1, H_α, R_α, and L_α represent the micellar isotropic, hexagonal, rectangular, and lamellar phases, respectively.

hand, Dietrich and co-workers [38] recently reported smaller values of the twist elastic constant (k_{22}) in a mixture of N, N-dimethyl-N-ethylhexadecylammonium bromide (CDEAB)/decanol/ water, in the nematic phase with respect to the other constants. They found $k_{11} = (1.54 \pm 0.08)\,pN$, $k_{22} = (0.300 \pm 0.018)\,pN$, and $k_{33} = (20.8 \pm 1.3)\,pN$. Assuming that these results extend to other lyotropic nematics, an immediate conclusion is that the single-constant approximation commonly used to simplify the writing of the elastic free energy [39], does not hold.

Despite the fact that the nematic phase is usually encountered in a small region of the triangular phase diagram of ternary mixtures [40] (e.g., Figure 5.10), the number of mixtures presenting the three nematic phases is big and, taking into account the 'recipe' described in [33], other mixtures can be easily prepared.

References

[1] A. M. Figueiredo Neto and S. R. A. Salinas (2005). *The Physics of Lyotropic Liquid Crystals, Phase Transitions and Structural Properties.* Oxford: Oxford University Press.

[2] D. Markovitsi, A. Mathis, J. Simon, J. C. Wittman, and J. Le Moigne (1980). Annelides V: A new type of lyotropic mesomorphic phase. *Mol. Cryst. Liq. Cryst.*, 64, 121–125.

[3] L. L. Ma, Q. Tang, M. X. Liu, X. Y. Liu, J. Y. Liu, Z. L. Lu, Y. G. Gao, and R. Wang (2020). [12]aneN3-based gemini-type amphiphiles with two-photon absorption properties for enhanced nonviral gene delivery and bioimaging. *ACS Appl. Mater. Interfaces*, 12, 40094–40107.

[4] F. M. Menger and J. Ding (1996). Spiro-Tenside und -Phospholipide: Synthese und Eigenschaften. *Angew. Chem.*, 108, 2266–2268.

[5] Y. Zhao (2007). Facial amphiphiles in molecular recognition: From unusual aggregates to solvophobically driven foldamers. *Current Opinion in Colloid & Interface Science*, 12, 92–97.

[6] A. S. Al-Shereiqi, B. J. Boyd, and K. Saito (2015). Photo-responsive self-assemblies based on bio-inspired DNA-base containing bolaamphiphiles. *Chemical Communications*, 51, 5460–5462.

[7] L. L. Schramm, E. N. Stasiuk, and D. G. Marangoni (2003). Surfactants and their applications. *Annu Rep Prog Chem Sect C Phys Chem*, 99, 3–48.

[8] K. Holmberg, B. Jonsson, B. Kronberg, and B. Lindman (2002). *Surfactants and polymers in aqueous solution*, 2nd edn. Chichester: Wiley.

[9] R. Nagarajan (2011). Amphiphilic surfactants and amphiphilic polymers: principles of molecular assembly. In: R. Nagarajan (ed.), *Amphiphiles: Molecular Assembly and Applications, ACS Symposium Series*, vol 1070, Washington, DC: American Chemical Society, chapter 1, pp 1–22.

[10] S. W. H. Shah, C. Schwieger, Z. Li, J. Kressler, and A. Blume (2017). Effect of perfluoroalkyl endgroups on the interactions of tri-block copolymers with monofluorinated F-DPPC monolayers. *Polymers*, 9, 555.

[11] K. Holmberg, B. Jönsson, B. Kronberg, and B. Lindman (2002). *Surfactants and Polymers in Aqueous Solution.* Chichester: John Wiley & Sons, Ltd.

[12] E. Guazzelli, F. Perondi, F. Criscitiello, C. Pretti, M. Oliva, V. Casu, F. Maniero, L. Gazzera, G. Galli, and E. Martinelli (2020). New amphiphilic copolymers for PDMS-based nanocomposite films with long-term marine antifouling performance. *J. Mater. Chem. B*, 8, 9764–9776.

[13] G. C. Feast, T. Lepitre, X. Mulet, C. E. Conn, O. E. Hutt, G. P. Savage, and C. J. Drummond (2014). The search for new amphiphiles: Synthesis of a modular, high-throughput library. *Beilstein J. Org. Chem.*, 10, 1578–1588.

[14] S. Öztürk, S. Okay, and A. Yıldırım (2020). Synthesis, anticorrosion, antibacterial, and antifungal activity of new amphiphilic compounds possessing quinazolin-4(3H)-one scaffold. *Russian Chemical Bulletin*, 69, 2205–2214.

[15] C. Tanford (1980). *The Hydrophobic Effect: Formation of Micelles and Biological membranes.* New York: Wiley.

[16] J. N. Israelachvili (1991). *Intermolecular and Surface Forces.* New York: Academic Press.

[17] R. J. Hunter (1993). *Foundations of Colloid Science*, vol 1, 5th edn. New York: Oxford Science Publications.

[18] N. A. Mazer, G. B. Benedek, and M. C. Carey (1976). An investigation of the micellar phase of sodium dodecyl sulfate in aqueous sodium chloride solutions using quasielastic light scattering spectroscopy. *J. Phys. Chem.*, 80, 1075–1085.

[19] K. Nesmeràk and I. Nemcova (2006). Determination of critical micelle concentration by electrochemical means. *Analytical Letters*, 39, 1023–1040.

[20] S. Deodhar, P. Rohilla, M. Manivannan, S. P. Thampi, and M. G. Basavaraj (2020). Robust method to determine critical micelle concentration via spreading oil drops on surfactant solutions. *Langmuir*, 36, 8100–8110.

[21] A. Angelova, B. Angelov, R. Mutafchieva, S. Lesieur, and P. Couvreur (2011). Self-assembled multicompartment liquid crystalline lipid carriers for protein, peptide, and nucleic acid drug delivery. *Accounts of Chemical Research*, 44, 147–156.

[22] Y. Huang and S. Gui (2018). Factors affecting the structure of lyotropic liquid crystals and the correlation between structure and drug diffusion. *RSC Adv.*, 8, 6978–6987.

[23] D. Roux, M. E. Cates, U. Oleson, R. C. Ball, F. Nallet, and A. M. Bellock (1990). Light scattering from a surfactant 'sponge' phase: Evidence for a hidden symmetry. *Europhys. Lett.*, 11, 229–234.

[24] M. Magalhães, A. M. Figueiredo Neto, and A. C. Tromba (2004). Bulk structure of the (inverted) sponge phase of the lyotropic mixture of sodium dodecyl sulfate, 1-pentanol, water, and cyclohexane. *J. Phys. Chem. B*, 108, 15962–15968.

[25] K. Fontell and L. Mandell (1993). Phase equilibria and phase structure in the ternary systems sodium or potassium octanoate-octanoic acid-water. *Colloid and Polymer Science*, 271, 974–991.

[26] P. Walde, T. Namani, K. Morigaki, and H. Hauser (2006). In: G. Gregoriadis (ed.), *Liposome Technology, Vol. I*, 3rd edn., New York: Informa Healthcare, pp 1–19.

[27] I. Dierking and A. M. Figueiredo Neto (2020). Novel trends in lyotropic liquid crystals. *Crystals*, 10, 604.

[28] L. J. Yu and A. Saupe (1980). Observation of a biaxial nematic phase in potassium laurate-1-decanol-water mixtures. *Phys. Rev. Lett.*, 45, 1000–1003.

[29] M. J. Freiser (1970). Ordered states of a nematic liquid. *Phys. Rev. Lett.*, 24, 1041–1043.

[30] R. Alben (1973). Phase transitions in a fluid of biaxial particles. *Phys. Rev. Lett.*, 30, 778–781.

[31] J. C. Tolédano and P. Tolédano (1987). *The Landau Theory of Phase Transitions.* Singapore: World Scientific, chapter 4.

[32] K. D. Lawson and T. J. Flautt (1967). Magnetically oriented lyotropic liquid crystalline phases. *J. Am. Chem. Soc.*, 89, 5489–5491.

[33] E. Akpinar and A. M. Figueiredo Neto (2019). Experimental conditions for the stabilization of the lyotropic biaxial nematic mesophase. *Crystals*, 9, 158.

[34] Y. Hendrikx, J. Charvolin, and M. Rawiso (1986). Uniaxial-biaxial phase transition in lyotropic nematic solutions: Local biaxiality in the uniaxial phase. *Phys. Rev. B*, 33, 3534–3537.

[35] Y. Galerne, A. M. Figueiredo Neto, and L. Liébert (1987). Microscopical structure of the uniaxial and biaxial lyotropic nematics. *J. Chem. Phys.*, 87, 1851–1856.

[36] Y. Galerne and J. P. Marcerou (1983). Temperature behavior of the order-parameter invariants in the uniaxial and biaxial nematic phases of a lyotropic liquid crystal. *Phys. Rev. Lett.*, 51, 2109–2111.

[37] T. Haven, D. Armitage, and A. Saupe (1981). Bend and splay elastic constants and the rotational viscosity of nematic decylammonium chloride and ammonium chloride water mixtures. *J. Chem. Phys.*, 75, 352–355.

[38] C. F. Dietrich, P. J. Collings, T. Sottmanna, P. Rudquistd, and F. Giesselmann (2020). Extremely small twist elastic constants in lyotropic nematic liquid crystals. *PNAS*, 117, 27238–27244.

[39] P. Sheng (1976). In: E. B. Priestley, P. I. Wojtowicz, and P. Sheng (eds), *Introduction to Liquid Crystals*, . New York: Plenum Press.

[40] Y. Hendrikx, J. Charvolin, M. Rawiso, L. Liébert, and M. C. Holmes (1983). Anisotropic aggregates of amphiphilic molecules in lyotropic nematic phases. *J. Phys. Chem.*, 87, 3991–3999.

6
Chiral lyotropic liquid crystals

6.1 Introduction

In this chapter the chiral mesophases of lyotropic mixtures will be presented. We start with the cholesteric lyotropic liquid crystals, obtained by doping nematic mesophases with chiral molecules or using, directly, chiral amphiphilic molecules. Lyotropic blue phases and the recently observed chiral lamellar structure, with its remarkable features, will also be discussed.

6.2 Micellar cholesteric lyotropic liquid crystals (MCLC)

In 1975 Tracey and Diehl reported NMR results from a lyotropic mixture of sodium n-decylsulfate/sodium decyl-2-sulfate/sodium sulfate/n-decanol/D-alanine/D_2O [1]. Their results may suggest that alanine molecules could penetrate into the amphiphilic molecular aggregates giving rise to more complex interaction between these aggregates. The systematic research with MCLC started with the work of Radley and Saupe [2], who prepared mixtures of caesium decylsulfate/decanol/water/d-tartaric acid; ammonium decylsulfate/ammonium sulfate/decanol/water/brucine sulfate(BS) and decylammonium chloride/ammonium chloride/water/cholesterol, where the last component of each mixture is the optically active compound. This type of mixture, where an optically active molecule is incorporated into the usual lyotropic mixture, stabilizes the *induced* (or, *extrinsic*) cholesteric mesophases. A phenomenological model considering the elastic twist deformation of the original nematic structure (in the framework of the Frank–Oseen elastic theory [3]) was proposed to relate the cholesteric helical pitch, P, with the concentration of chiral molecules present in the mixture, c_m, in moles of solute/moles of solution, shown in Eq. 6.1:

$$\frac{1}{P} = \alpha c_m, \tag{6.1}$$

where α depends on the elastic constants of the medium [2]. In the limit of $c_m \to 0$ and, consequently, $P \to \infty$, α is defined as the *Helical Twisting Power* (HTP—dimension in SI, m^{-1}) of the mesophase [4,5].

In 1980, Acimis and Reeves reported the existence of an *intrinsic* cholesteric lyotropic mesophase, identified in a mixture of (d or l) α-alanine hydrochloride decylester/Na_2SO_4/D_2O [6]. In this mixture, the (d or l) α-alanine hydrochloride

Lyotropic Liquid Crystals. Ingo Dierking and Antônio Martins Figueiredo Neto, Oxford University Press.
 DOI: 10.1093/9780191924927.003.0006

decylester is an optically resolved amphiphile. When the mixture is prepared with both D- and L-enantiomers in the same concentrations, a nematic phase may be stabilized. Mesophases were identified by direct optical microscopy observation, with the appearance of the characteristic fingerprint textures, and nuclear magnetic resonance spectra.

As in nematics, three types of phases were identified in MCLC. These phases originate from the two uniaxial and the biaxial nematic phases, doped with a chiral agent. These mesophases were named calamitic cholesteric, Ch_C, discotic cholesteric, Ch_D, and biaxial cholesteric, Ch_B [7,8]. Figure 6.1 shows the typical textures, observed under a polarized optical microscope, of the Ch_C (Figure 6.1(a)), Ch_D (Figure 6.1(b)), and Ch_B (Figure 6.1(c)). Let us discuss the case where the nematic mesophases are obtained from mixtures where the main amphiphile has carbonic chains. The anisotropy of diamagnetic susceptibility, ($\Delta\chi = \chi_{\parallel} - \chi_{\perp}$, where

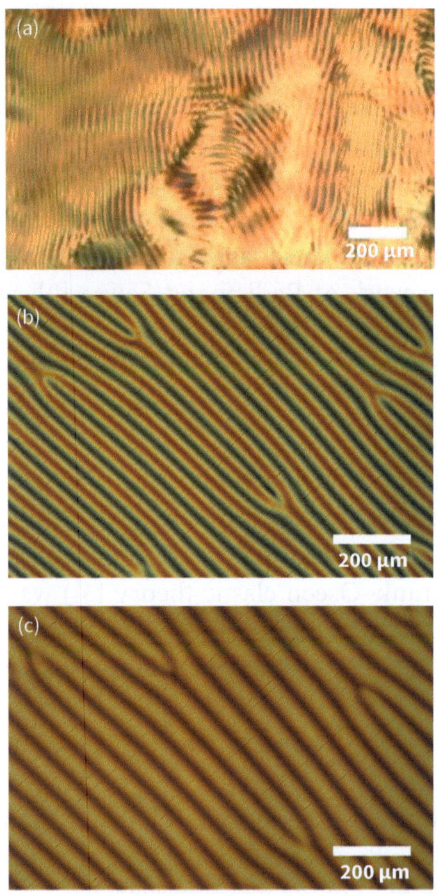

Figure 6.1 Textures of the lyotropic cholesteric mesophases under a polarized optical microscope. Mixture of potassium laurate/K_2SO_4/dodecanol/water/brucine sulfate: (a) Ch_C phase, temperature 25.0 °C; Mixture of potassium laurate/l-decanol/water/brucine sulfate: (b) Ch_D phase, temperature 18.9 °C; (c) Ch_B phase, temperature 20.0 °C.

$\chi_{\|}$ and $\chi_{\perp}$ are the susceptibilities parallel and perpendicular to the director of the phases) of the molecular arrangement is responsible for the alignment of the micelles in an external magnetic field. The N_C phase has $\Delta\chi > 0$ and N_D $\Delta\chi < 0$, and will align the mesophase director parallel and perpendicular to the field, respectively [9]. On the other hand, perfluorated amphiphiles employed as the main amphiphile in the mixture were shown to stabilize mesophases where the signs of $\Delta\chi$ are inverted with respect to those of the carbonated amphiphile mesophases [10].

Coming back to the carbonated-chain amphiphiles as the main component of the mixture that present cholesteric mesophases, in the presence of an external magnetic field, $\vec{H}$, Ch_D (Figure 6.1(a)) and Ch_B (Figure 6.1(b)) phases align the helical helix parallel to $\vec{H}$, and the cholesteric structure is unwound in the Ch_C phase. A tip for experimentalists to obtain the fingerprint texture of the Ch_C phase in the polarized optical microscope is: apply the magnetic field until the texture reaches a perfect planar alignment of the director parallel to $\vec{H}$, after that, leave the sample to relax in the absence of $\vec{H}$; this relaxation process may take hours, depending on the viscosity of the medium, concentration of chiral agent, and temperature, among others; the fingerprint texture is spontaneously formed (Figure 6.1(c)).

A striking feature observed in the fingerprint textures of cholesteric phases, in the Ch_D to Ch_B transition, is the disappearance of the homeotropic regions [7]. In the Ch_D phase, there are regions where the director (twisted in a helicoidal arrangement) is aligned along the light-beam propagation direction in the polarized optical microscope. The local director orients perpendicularly to the helical axis, forming a plane that twists along the helical axis. These regions are homeotropic, even under the action of an external magnetic field, and characterize the Ch_D phase. These regions are no longer present in the Ch_B phase, since the two optical axes of the Ch_D phase form an angle $\pm\theta/2$ with respect to the normal direction to the helical axis, but still in the same plane defined in the Ch_B phase. In this particular condition, no homeotropic regions are observed in the polarized optical microscope. The accurate observation of these homeotropic regions and interferometric measurements of the chirality-induced biaxiality in the vicinity of the Ch_D to Ch_B transition, allows study of the behaviour of the order parameter that characterizes the transition [11,12]. In the framework of the Landau–de Gennes model, the free-energy density of the cholesteric phase, as a function of the temperature, T, may be written as:

$$F = \frac{1}{2}a(T)\zeta^2 + \frac{1}{4}b\zeta^4 - cq^2\zeta, \tag{6.2}$$

where $a(T) = a_0(T - T_C)$, T_C is the uniaxial to biaxial transition temperature, a_0, b, and c are the Landau coefficients, ζ is the scalar order parameter that characterizes the biaxial phase, $q = 2\pi/P$, and P representing the helical pitch. The coefficient c is of the order of magnitude of the Frank elastic constant. The last term in Eq. 6.2 describes the contribution of the gradients in the free energy, coupling the scalar order parameter, ζ, to the chirality field, q^2. To the first order, the induced biaxial order parameter in the Ch_D phase obtained from minimization of Eq. 6.2 with respect to ζ, is written as:

$$\zeta \cong \frac{cq^2}{a_0 (T - T_C)}. \tag{6.3}$$

Inspecting the details of Eq. 6.3, and its minimization with respect to ζ, in the vicinity of T_C we can observe the pretransitional chirality-induced biaxiality in the Ch_D phase [10]. The maximum limit of this biaxiality at $T = T_C$ is:

$$\zeta \sim \left(\frac{cq^2}{b}\right)^{1/3}. \tag{6.4}$$

The asymptotic behaviours for ζ (far from T_C) is $\zeta = 0$ in the Ch_D and $\zeta = -\left(\frac{a_0}{b}\right)(T - T_c)$ in the Ch_B phase.

From the experimental point of view, it is possible to investigate the characteristic of the Ch_D to Ch_B phase transition measuring the optical birefringences in magnetically oriented samples in the shape of a slab between crossed polarizers [10]. The external magnetic field is oriented along the 1 axis of the laboratory frame and the light beam is along the 3 axis. With the measurements of the birefringences $\Delta n = n_3 - n_2$ and $\delta n = n_2 - n_1$ as a function of the temperature, one can access the biaxial order parameter $\zeta \sim \delta n / \Delta n$. Figure 6.2 shows ζ^2 as a function of the temperature around the Ch_D to Ch_B transition of a lyotropic mixture of potassium laurate(KL)/l-decanol(DeOH)/H_2O/1-N-lauroyl potassium alaninate (1-LAK). From the asymptotic behaviour of ζ^2, the fit to the experimental results allows the determination of $|a_0|/b$ and $cq^2/|a_0|$. As $q = 2\pi/P$ and the helical pitch depends on the concentration of the chiral agent in the mixture, it is possible to measure the ratio $cq^2/|a_0|$ for mixtures with cholesteric phases with different pitches, i.e., different q. Figure 6.3 shows $cq^2/|a_0|$ as a function of q^2 of mixtures of KL/DeOH/H_2O, doped with different

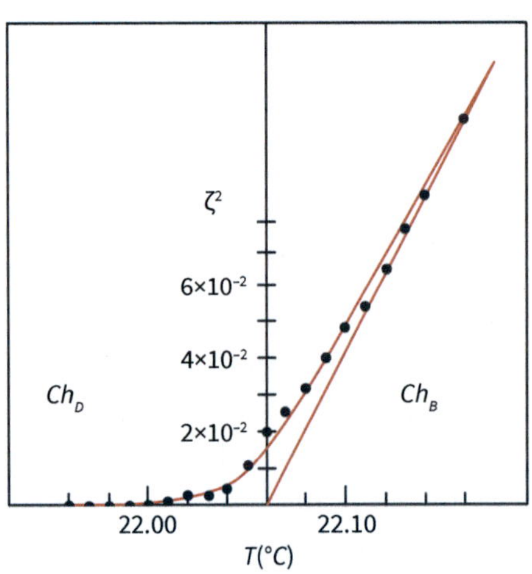

Figure 6.2 Squared biaxial order parameter vs temperature, across the Ch_D to Ch_B transition of a l-LAK doped sample. The solid line is a fit to the Landau-type behaviour.

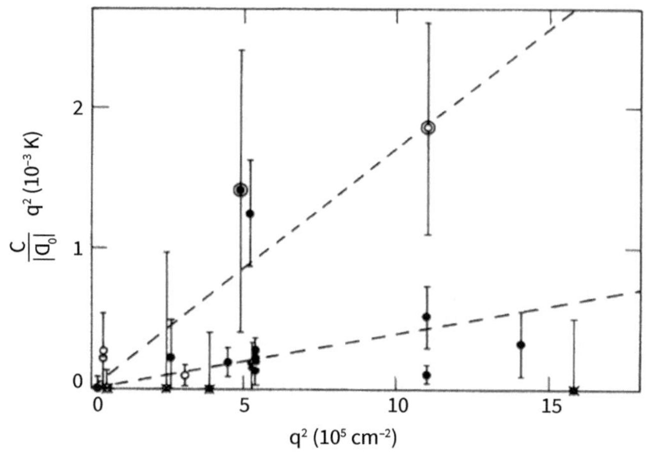

Figure 6.3 Depiction of the quantity $cq^2/|a_0|$ as a function of q^2 of mixtures of KL/DeOH/H_2O, doped with different concentrations of one of the three chiral agents: l-LAK, BS, and d-octanol [10].

concentrations of one of the three chiral agents: l-LAK, BS, and d-octanol [10]. These values allow the evaluation of the bare correlation length $\xi_0 \sim 15$ nm, showing that the correlation length $\xi = \xi_0[(T - T_C)/T_C]^{-1/2}$ is much larger than the micellar dimensions. Inspection of Figure 6.2 reveals that the chirality field renders the Ch_D to Ch_B transition continuous, as described theoretically by Brand and Pleiner [13]. The Ch_D to Ch_B transition does not cross transition lines in phase diagrams, similar to the liquid–gas transition after the critical point. Measurements of ζ in the temperature range around T_C of $\Delta T/T_C \sim 10^{-4}$ show that the critical exponents for the order parameter, β, and for the susceptibility, γ, agree with the theoretical results for the XY model, $\beta = 0.38$ and $\gamma = 1.35$, respectively [12].

An interesting question in doped (extrinsic) cholesteric lyomesophases is the effect of the chiral molecule on the micelles and their local ordering. To investigate this issue, Small-Angle X-Ray Scattering (SAXS) experiments and an advanced modelling analysis [14] were performed in a magnetically aligned calamitic nematic phase (N_C) and an unwound Ch_C phase [15]. The lyotropic mixture that shows the N_C phase is composed of KL/potassium sulfate/dodecanol/water and, to obtain the Ch_C phase, brucine sulfate (BS) was added to the mixture. Samples were subjected to an external magnetic field (axis 1 of the laboratory frame), enough to align the director of the N_C phase, and unwind the Ch_C phase. In both conditions, the larger size of the micelles is oriented along axis 1. Figure 6.4 shows the SAXS results of the N_C phase (a) and Ch_C phase (b), along axis 1 (blue dots) and axis 3 (red dots). The advanced modelling analysis allows the investigation of the structure and local ordering in complex fluids by the full scattering-curve fitting of the SAXS data. An advantage with respect to the usual methods is the separation of the (micellar) form factor from the structure factor (pseudo-lamellar ordering—see e.g., Chapter 5). The solid lines in Figure 6.4 represent the fits of the model to the experimental data, in the entire range of the momentum transfer modulus, $q = (4\pi/\lambda)\ sin\theta$, where 2θ is the scattering angle and λ is the X-ray wavelength. The SAXS results revealed

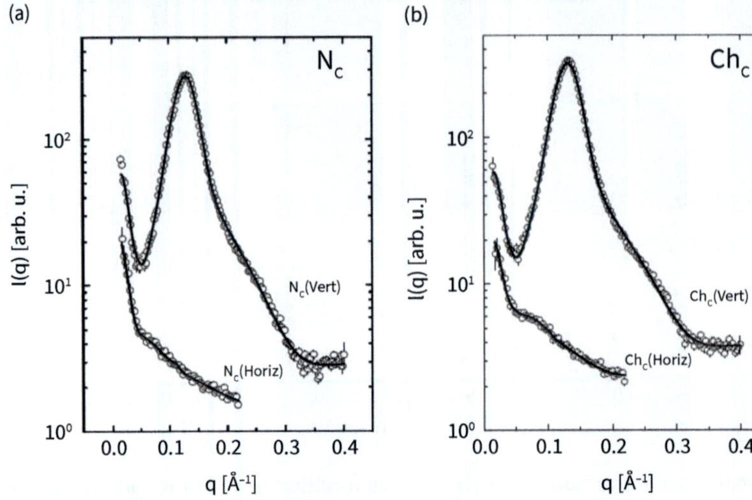

Figure 6.4 Scattered intensity as a function of q along the axes 1 and 3 for the oriented (a) N_C. and (b) unwound Ch_C phases. Open circles: Experimental data. Solid lines: Model fits. (Reproduced with permission from ref.[15]).

that the shape and shape anisotropy of the micelles encountered in the N_C phase is not modified by the doping with BS. On the other hand, the BS molecules located between micelles in the Ch_C phase (the amount of BS molecules in the structure was evaluated to about two molecules per micelle) imposes a higher positional correlation between micelles along the direction of the pseudo-lamellar ordering. The orientational order parameter measured in the N_C phase is higher than that of the Ch_C phase, due to the presence of the BS molecule in the structure [15].

6.3 Lyotropic blue phases (BP)

The existence of the blue phases in lyotropic liquid crystals is still a matter of discussion. In 1983, Lee and Labes [16] reported results obtained with the lyotropic mixture of disodium cromoglycate/allo-4hydroxy-l-proline/water, identifying a cholesteric and an isotropic phase in the partial phase diagram. Slowly cooling down the mixture from the isotropic to the cholesteric phase, just below the clearing point, a coexistence between the isotropic and a blue phase, characterized by the presence of platelets in the textures, was claimed by authors. The association between the presence of a platelet texture and the blue phase was suggested by Coates and Gray in 1973 [17], observing the thermotropic behaviour of textures of different cholesterol derivatives. However, after a reanalysis of the experimental results, still in 1983, this observation of a blue phase was retracted [18].

Blue phases were identified in lyotropic mixtures of DNA/water, with DNA fragments of length in the range from 50 nm to 3000 nm [19]. BP were found in small temperature ranges between the isotropic and cholesteric mesophases, in a narrow range of relative concentrations of the components of the mixture. The transition from the isotropic to the BP is of first order. The experimental technique employed to characterize the BP is, mainly, electron microscopy of freeze-fracture replicas. The 'basic unit' in the BP structures are concentric cylinders, where the DNA molecules twist along a given direction (see, e.g., Figure 6.5(a)). In the BP-III phase, or 'Blue Fog', the cylinders are globally disordered in the water, but keeping some ordering in the cylinders' length scale (Figure 6.5(b)). Electron microscopy results show that the cylinders do not occupy all the volume of the sample, evidencing an isotropic distributed clustering of them in the structure [19]. A more compact arrangement of cylinders, organized in a cubic structure, was also observed. The basic units occupy the whole space in the electron microscopy texture, corresponding either to the BP-I or BP-II phase (Figures 6.5(c),(d)). An interesting observation in the lyotropic BP phases is that, differently from the BP of thermotropic liquid crystals, the helical pitch is much smaller than the one observed in the lyotropic cholesteric phases. Figure 6.6 shows the electron microscopy of freeze-fracture replicas of the BP-III phase (Figure 6.6(a)) and the BP-I or BP-II phase (Figure 6.6(b)), with the characteristics described above.

In 1995, K. Radley [20] identified a mesophase, located between the isotropic and cholesteric calamitic phase, named I_C. The amphiphilic lyotropic mixture investigated was composed of N-tetradecanoyl-L-alaninate/decanol/CsCl/water. From experimental results obtained with polarizing optical microscopy, laser diffraction, and nuclear magnetic spectroscopy, a partial phase diagram was stablished. The I_C phase lies in temperature ranges as large as about 40 °C. It is optically isotropic and the reflected wavelength is in the infrared region, indicating a larger periodic structure (periodicity in the range of micrometres). These two characteristics differ from those observed in thermotropic BPs. Despite these facts, Radley suggests that the I_C phase may be a lyotropic BP of cubic structure, where cylindrical micelles are

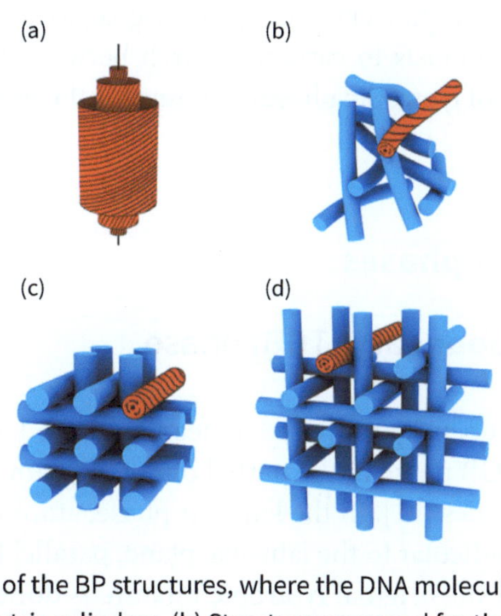

Figure 6.5 (a) Basic unit of the BP structures, where the DNA molecules twist along a given direction, forming concentric cylinders. (b) Structure proposed for the BP-III. In the case of the BP-II and BP-I, sketches of the simple cubic (c) and face-centred (d) structures, respectively.

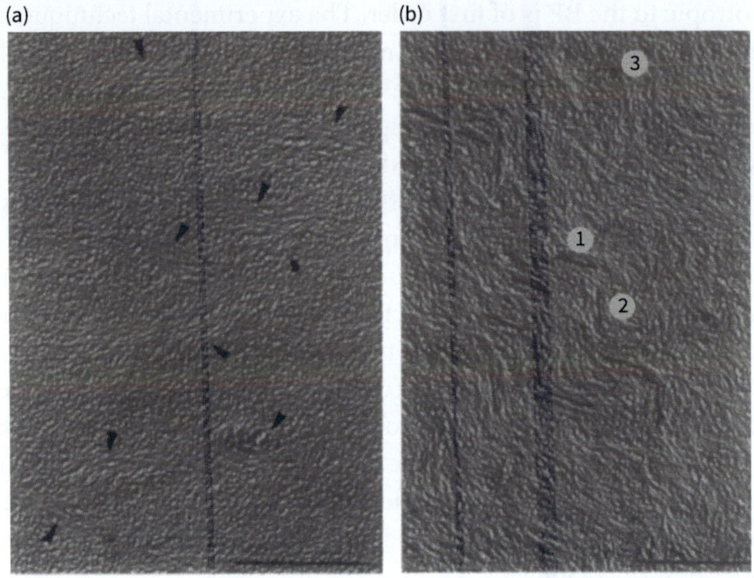

Figure 6.6 Lyotropic mixture of DNA/water. Electron microscopy of freeze-fracture replicas in the blue phase mesophases. Scale bars = 0.5 μm. (a) Blue phase BP-III, DNA fragments of 50 nm, bundles (arrowheads) dispersed in the isotropic phase; (b) blue phase BP-I or BP-II, with the bundles oriented in different directions (1,2,3).
(Reproduced with permission from ref.[19]).

arranged in a structure similar to those of Figs. 6.5c,d. Water fills the intermicellar voids of the cubic structure. However, a more detailed characterization of the I_C phase, with complementary experimental techniques, is necessary to identify it as an amphiphilic lyotropic BP.

In summary, except in the case of the DNA/water mixture, the existence of BP in amphiphilic lyotropic liquid crystals is still a goal to be pursued. It remains a challenge to experimentalists to carefully search between the isotropic and the cholesteric mesophases of lyotropic phase diagrams for the existence of the remarkable BP.

6.4 Chiral lamellar phases

6.4.1 Twist-grain boundary (TGB) phase

In 1997, Kamien and Lubensky proposed a mean-field model to describe the possible appearance of a TGB phase in the partial phase diagram of lyotropic mixtures containing chiral molecules [21]. In the lamellar phase, amphiphilic molecules are, mainly, oriented perpendicular to the lamellae plane, parallel to each other. If these molecules are chiral, or the original lyotropic mixture is doped with chiral agents, there is a tendency of the structure, and consequently the molecules, to twist. These

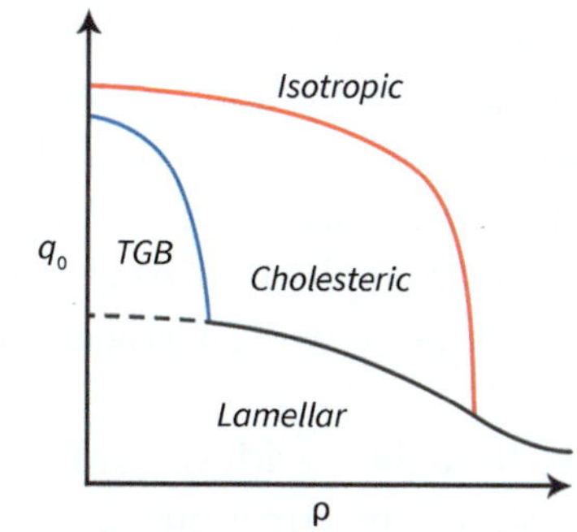

Figure 6.7 Partial phase diagram theoretically predicted for a lyotropic mixture of chiral amphiphilic molecules and a solvent. ρ and q_0 represent the number density of amphiphilic molecules and the chirality factor.

competing tendencies may lead to the stabilization of a TGB phase, at least in principle. The partial phase diagram predicted by the theory is shown in Figure 6.7. The numerical density of amphiphilic molecules is represented by ρ and the q_0 is the chirality factor. Screw dislocations, with the centre occupied by the solvent, would be formed in the structure. The periodic arrangement of screw dislocations would form the TGB phase.

From the experimental point of view, it was only in 2011 when a report claiming the observation of the TGB phase in a lyotropic mixture of graphene oxide and water was published [22]. The process to allow the graphene sheets to be dispersed in water involves the attachment of hydroxyl, carboxyl, and epoxide groups to them. Sheets have an average thickness of 0.8 nm and an average lateral width of 0.81 μm. The volume fraction of sheets in the region of the partial phase diagram where the TGB phase would have been identified is about 0.6%. An aspect that is still unclear in this experiment is the origin of the chirality that seems to be observed. Different from the local symmetry of banana-like molecules, the lamellar-type packing of the sheets is uniaxial, that avoids the helical macroscopic structure. If it were possible to dope with (or attach to the sheets) a chiral molecule, one could overcome this difficulty in interpreting the experimental data.

It is clear that graphene oxide/water is a very interesting system where not only the liquid crystalline packing of the sheets brings new and interesting questions, but also because of its potential technological applications due to its mechanical and electrical properties.

6.4.2 Tilted phase and ferroelectricity

In 1975 Bob Meyer and co-workers published a seminal paper about the existence of ferroelectric liquid crystals [23]. Symmetry arguments were employed to predict that chiral molecules organized in smectic C structure may give rise to a ferroelectric liquid crystalline phase. The outstanding group of chemists in Orsay (L. Liébert, L. Strzelecki, and P. Keller), with the collaboration of Meyer, succeeded in synthesizing the p-decyloxybenzylidene-p′-amino2-methylbutyl cinnamate (DOBAMBC). In their experiments, it was possible to demonstrate the presence of a spontaneous polarization, easily oriented by an applied external field. Since then, the search for new liquid crystalline materials showing ferroelectric behaviour started, and continues to this day.

The symmetry arguments brought by Meyer are so strong that it would be expected that this behaviour should be encountered not only in thermotropic liquid crystals, but also in lyotropics. The advances in chemical synthesis and the focus on new molecules (surfactants) that interact with neighbours in a particular way to avoid full rotations of them around their principal cylindrical axis opened the path to encounter the lyotropic smectic C (SmC) [24]. These surfactants are constituted by a rigid rod-like hydrophilic moiety of variable length. At both ends, monodisperse oligo- (ethylene glycol) units are attached. Binary mixtures of this type of molecule in water showed a rich phase diagram, including the SmC phase. Another example of the SmC mesophase in lyotropics was reported by Ujiie and Yano [25] in mixtures of ionic liquid crystalline amphiphiles {hydrophilic poly(ethyleneimine) unit with hydroxy side-groups, attached to an aromatic mesogenic core}/water.

Chemists continued the synthesis of more molecules, searching for those that could fulfil the local molecule–molecule interaction to stabilize Sm-type structures and, moreover, the inter-layer coupling that could favour a chiral arrangement, equivalent to the thermotropic SmC* phase. In 2013 the first report about a lyotropic SmC* phase in a partial phase diagram was published [26]. The molecule (named diol 1), that was designed to present both thermotropic and lyotropic characteristics, is shown in Figure 6.8. The lyotropic system is composed of the mixture of diol 1/water (or formamide). The phase diagram of the mixture is shown in Figure 6.9, where a rich polymorphism is observed, including the SmC* mesophase. X-ray diffraction clearly shows the presence of a SmC phase, with the diffraction peaks associated with the lamellar-like structure, coexisting with the peaks showing the molecular tilt with respect to the normal to the layers. Figure 6.10 shows optical microscopic textures of the lyotropic SmC* phase: Mixture diol 1/water, a) Schlierentexture, b) broken fan texture; Mixture diol 1/formamide, c) and d) surface-stabilized states with domains in opposite tilt directions. The SmC*

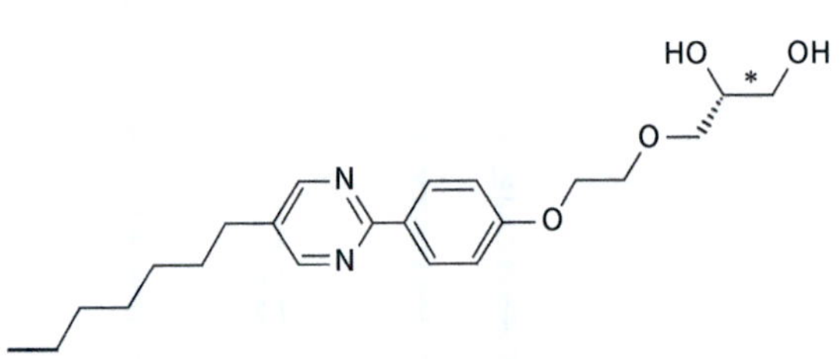

Figure 6.8 Chemical structure of the diol 1 molecule.

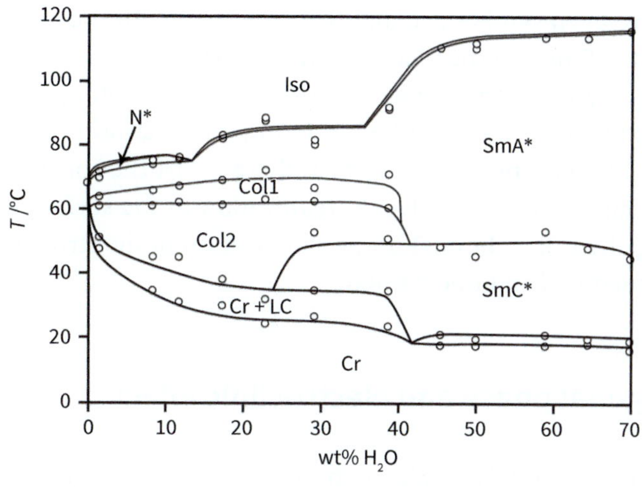

Figure 6.9 Phase diagram of the mixture diol 1/water.

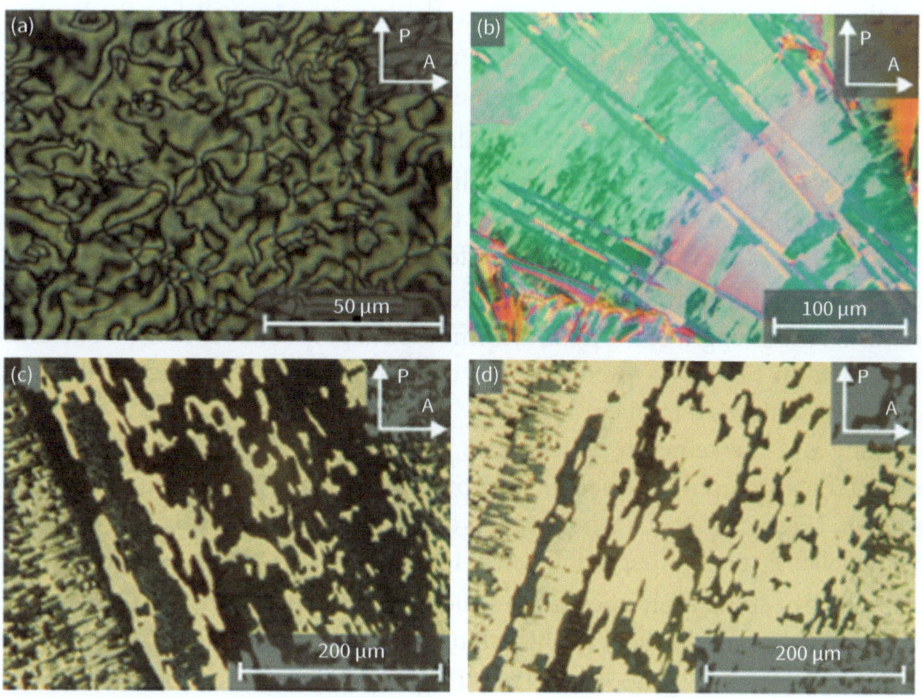

Figure 6.10 Optical microscopic textures of the lyotropic SmC* phase: Mixture diol 1/water, (a) smectic Schlierentexture, (b) broken fan texture; mixture diol 1/formamide, (c) and (d) surface-stabilized states with domains in opposite tilt directions. P and A represent the polarizer and the analyser.
(Reproduced with permission from ref.[26]).

characteristic was demonstrated since the mesophase showed spontaneous polarization. Two stable surface-stabilized states were observed switching the sense of the electric field applied to the sample (Figures 6.10(c) and (d)). More recently, new amphiphiles have been synthesized and their mixtures with polar solvents showed the presence of the SmC* in the phase diagrams [27].

An intriguing question concerning the existence of the SmC* phase in lyotropics remains to be explained: how the tilted ordering information is 'transmitted' from one layer to the others, since between them there is just the solvent (e.g., water or formamide)? Claims about the solvent organization in the surface of the lamellae and its role (e.g., hydrogen bonds) in the stabilization of the structure still need to be clearly demonstrated and understood.

6.4.3 Orthogonal phase and electroclinic effect

Thermotropic liquid crystals, made of chiral molecules, may stabilize the smectic A mesophase (SmA*). In this phase, molecules are stacked in layers and are free to rotate along their major axis, giving the structure a uniaxial character, $\vec{k}$ and $\vec{n}$ being the unitary vectors perpendicular to the layers and parallel to the long axis of the molecules, respectively. In the absence of any external electric field, $\vec{k}$ and $\vec{n}$ are parallel to each other. However, if an external electric field, $\vec{E}$, is applied along the direction of the layers, i.e., perpendicular to $\vec{k}$, the uniaxial symmetry is broken, and a transverse polarization, $\overrightarrow{\delta P}$, is established, perpendicular to the $\vec{k} - \vec{n}$ plane, [28,29]. The effect of the electric field is to hinder the free rotation of the chiral molecules around their long axes. This effect, named 'electroclinic effect', depends on the strength of the applied electric field, E. An angle $\delta\theta$ between $\vec{k}$ and $\vec{n}$, that depends on the strength of the field, is formed. $\delta\theta$ is proportional to E, and inversely proportional to $(T - T_C^*)$, where T is the temperature and T_C^* the critical SmA* - SmC* phase-transition temperature.

Harjung and Giesselmann [30], in 2018, reported the phase diagram of the mixture G10 (an amphiphilic molecule, Figure 6.11)/formamide, showing a rich polymorphism (Figure 6.12). This molecule is chiral and showed the SmC* mesophase and a lamellar L_α^* phase, that is equivalent to the thermotropic SmA* (the * refers to the existence of chiral molecules in the structure). The unit vectors $\vec{k}$ and $\vec{n}$ lie in the y–z plane and the electric field applied along the x axis leads to the appearance of an electric polarization $\overrightarrow{\delta P}$ proportional to E (Figure 6.13). The electroclinic effect was observed in the L_α^* phase, in the vicinity of the SmC* mesophase. A square-wave

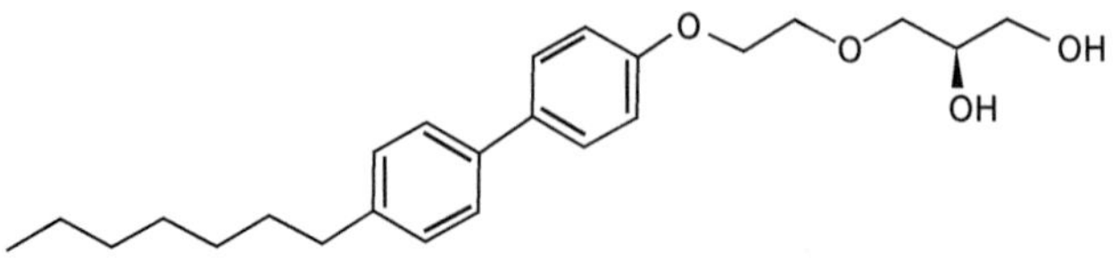

Figure 6.11 Molecular structure of the amphiphile G10. Mixtures of G10/formamide showed the lamellar L_α^* mesophase [30].

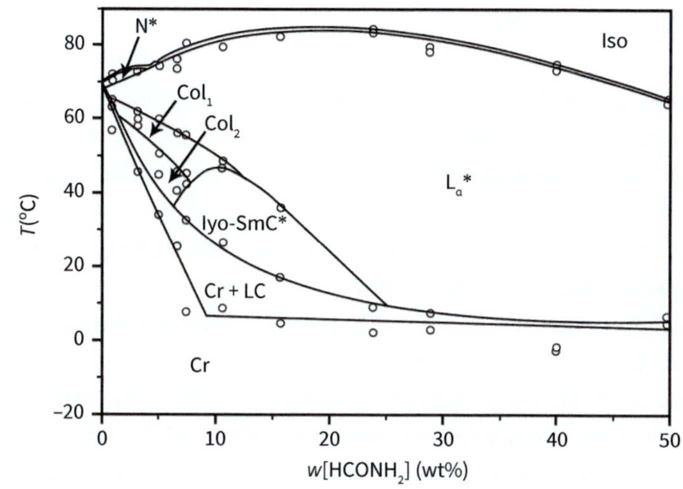

Figure 6.12 Phase diagram of the mixture G10/formamide. The lyotropic SmC* and the L_α^* are present.

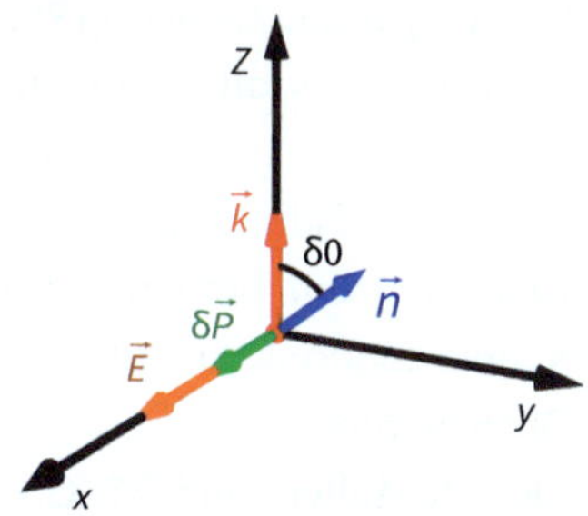

Figure 6.13 Sketch of the axes of the laboratory frame. The unit vectors $\vec{k}$ and $\vec{n}$ lie in the y–z plane. The electric field, applied along the x axis, leads to the appearance of an electric polarization $\vec{\delta P}$.

electric field, at a frequency of 400 Hz and amplitude of 1.9 V/μm, applied to the sample conveniently aligned in a polarized optical microscope, showed transmittance that depends on the sign of the electric field. The maximum value of the angle between $\vec{k}$ and $\vec{n}$ observed in these experimental conditions was $\delta\theta \sim 0.7$ *deg*, about one order of magnitude smaller than that obtained in thermotropic liquid crystals, under the same conditions.

The application of external electric fields in lyotropic liquid crystals is a tricky issue, since ions are present in the mixture and their role in the stabilization of a particular mesophase is essential. Low-frequency fields may destabilize a mesophase, and the minimum frequency, f_{min}, and amplitude, E_{max}, depend on the characteristics of the mixture, in particular the concentrations of ions present, viscosity of the solvent, and ions' mobility. In the absence of a reliable model to describe the role of these ions in the mixture, stability as a function of external electric fields, f_{min} and E_{max} have to be encountered experimentally.

References

[1] A. S. Tracey and P. Diehl (1975). Interaction of D-alanine and L-alanine with an optically-active model membrane system. *FEBS. Lett.*, 59, 131–132.

[2] K. Radley and A. Saupe (1978). Cholesteric states of micellar solutions. *Mol. Phys.*, 35, 1405–1412.

[3] P. Sheng (1976). In: E. B. Priestley, P. J. Wojtowicz, and P. Sheng (eds), *Introduction to Liquid Crystals*, New York: Plenum Press.

[4] J. Adams and W. Haas (1975). Pitch dependence on composition in mixtures of liquid crystals. *Mol. Cryst. Liq. Cryst.*, 30, 1–9.

[5] G. Gottarelli, M. Hibert, B. Samori, G. Solladie, G. P. Spada, and R. Zimmermann (1983). Induction of the cholesteric mesophase in nematic liquid crystals: Mechanism and application to the determination of bridged biaryl configurations. *J. Am. Chem. Soc.*, 105, 7318–7321.

[6] M. Acimis and L. W. Reeves (1980). A type I aqueous cholesteric lyomesophase. *Can. J. Chem.*, 58, 1533–1541.

[7] A. M. Figueiredo Neto, Y. Galerne, and L. Liebert (1985). Cholesterization of a biaxial nematic lyomesophase studied by x-ray diffraction and optical microscopy. *J. Phys. Chem.*, 89, 3939–3941.

[8] A. M. Figueiredo Neto and S. R. A. Salinas (2005). *The Physics of Lyotropic Liquid Crystals, Phase Transitions and Structural Properties.* Oxford: Oxford University Press.

[9] A. Saupe (1977). Textures, deformations, and structural order of liquid crystals. *J. Colloid and Interface Sci.*, 58, 549–558.

[10] N. Boden, P. H. Jackson, K. McMullen, and M. C. Holmes (1979). Are 'nematic' amphiphilic liquid crystalline mesophases thermodynamically stable? *Chem. Phys. Lett.*, 65, 476–479.

[11] T. Kroin, A. M. Figueiredo Neto, L. Liebert, and Y. Galerne (1989). Chirality-induced biaxiality at the uniaxial-to-biaxial cholesteric phase transition. *Phys. Rev. A*, 40, 4647–4651.

[12] Z. A. de Sant'Ana and A. M. Figueiredo Neto (1992). Uniaxial-to-biaxial cholesteric and nematic phase transitions on a lyotropic alcohol-free mixture. *Phys. Rev. A* 46, 7630–7635.

[13] H. R. Brand and H. Pleiner (1985). Cholesteric to cholesteric phase transitions in liquid crystals. *J. Phys. (Paris) Lett.*, 46, 711–718.

[14] C. L. P. Oliveira, B. B. Gerbelli, E. R. T. Silva, F. Nallet, L. Navailles, E. A. Oliveira, and J. S. Pedersen (2012). Gaussian deconvolution: A useful method for a form-free modeling of scattering data from mono- and multilayered planar systems. *J. Appl. Crystallography*, 45, 1278–1286.

[15] O. R. Santos, D. Reis, A. G. Oliveira-Filho, C. L. P. Oliveira, and A. M. Figueiredo Neto (2022). Structure and local order of lyotropic cholesteric calamitic phases: The effect of the chiral molecule. *Journal of Molecular Liquids*, 349, 118097.

[16] H. Lee and M. M. Labes (1983). Lyotropic cholesteric blue phase. *Mol. Cryst. Liq. Cryst.*, 82, 355–359.

[17] D. Coates and G. W. Gray (1973). Optical studies of the amorphous liquid-cholesteric liquid crystal transition: The 'blue phase'. *Physics Letters*, 45A, 115–116.

[18] M. Kuzma, H. Lee, and M. M. Labes (1983). A retraction of the report of a lyotropic cholesteric blue phase. *Mol. Cryst. Liq. Cryst.*, 92, 81–85.

[19] F. Livolant and A. Leforestier (1996). Condensed phases of DNA: Structures and phase transitions. *Prog. Polym. Sci.*, 21, 1115–1164.

[20] K. Radley (1995). An amphiphile system forming a lyotropic blue phase. *Liquid Crystals*, 18, 151–155.

[21] R. D. Kamien and T. C. Lubensky (1997). Chiral lyotropic liquid crystals: TGB phases and helicoidal structures. *J. Phys. II France*, 7, 157–163.

[22] Z. Xu and C. Gao (2011). Graphene chiral liquid crystals and macroscopic assembled fibres. *Nat. Commun.*, 2, 571.

[23] R. B. Meyer, L. Liébert, L. Strzelecki, and P. Keller (1975). Ferroelectric liquid crystals. *J. Phys. Lett. France*, 36, 69–71.

[24] M. A. Schafheutle and H. Finkelmann (1988). Shapes of micelles and molecular geometry synthesis and studies on the phase behaviour, surface tension and rheology of rigid rod-like surfactants in aqueous solutions. *Liquid Crystals*, 3, 1369–1386.

[25] S. Ujiie and Y. Yano (2000). Thermotropic and lyotropic behavior of novel amphiphilic liquid crystals having hydrophilic poly(ethyleneimine) units. *Chem. Commun.*, 1, 79–80.

[26] J. R. Bruckner, J. H. Porada, C. F. Dietrich, I. Dierking, and F. Giesselmann (2013). A lyotropic chiral smectic C liquid crystal with polar electrooptic switching. *Angew. Chem. Int. Ed.*, 52, 8934–8937.

[27] M. D. Harjung, C. P. J. Schubert, F. Knecht, J. H. Porada, R. P. Lemieux, and F. Giesselmann (2017). New amphiphilic materials showing the lyotropic analogue to the thermotropic smectic C* liquid crystal phase. *J. Mater. Chem.*, C 5, 7452–7457.

[28] S. Garoff and R. B. Meyer (1977). Electroclinic effect at the A-C phase change in a chiral smectic liquid crystal. *Phys. Rev. Lett.*, 38, 848–851.

[29] S. Garoff and R. B. Meyer (1979). Electroclinic effect at the A-C phase change in a chiral smectic liquid crystal. *Phys. Rev. A*, 19, 338–347.

[30] M. D. Harjung and F. Giesselmann (2018). Electroclinic effect in the chiral lamellar α phase of a lyotropic liquid crystal. *Phys. Rev. E*, 97, 032705.

7
Chromonics

7.1 Introduction

First things first: let us start with the denomination 'chromonics'. In the late 1970s/early 1980s, John Lydon, at that time a lecturer in Leeds (UK), was in contact with researchers that identified strange lyotropic mesophases in aqueous solutions of *Intal*. *Intal* is a trade name (*INT*erfere with *AL*lergies) of a drug in the UK, a bis-chromone derivative, used to treat asthma. In the United States of America, this drug received the trade name *Chromolyn*. The active drug is the molecule named *khellin* (Figure 7.1), that is essentially flat, formed by aromatic rings [1]. This type of molecule differs from those usually encountered in lyotropic liquid crystals (aliphatic molecules). Lydon proposed the name chromonic to this type of discovered lyotropic mixture, at that time. Since then, the nomenclature has been incorporated into the usual jargon of researchers to characterize lyotropic mixtures made of flat, aromatic molecules, with ionic parts on the periphery of the molecule.

Differently from molecular aggregates formed by means of the hydrophobic/hydrophilic interactions between aliphatic molecules, as encountered in lyotropic liquid crystalline mixtures, the primary units in chromonics are the aromatic molecules (disc- or plank-like in shape). The molecular packing is favoured by π-stacking interactions between flat molecules. They pile up in columns of different lengths. The forces that drive this face-to-face adhesion of molecules are of enthalpic origin. The further superstructure established with these columnar aggregates depends on the electric interactions between the surface ions of the stacks and entropic interactions. There is no evidence that the molecule–molecule interaction to form stacks depends on the number of molecules in solution. In other words, the stacking process is not a collective behaviour, as in the case of the micelle formation in aliphatic lyotropics, where a critical micellar concentration exists. Even at small

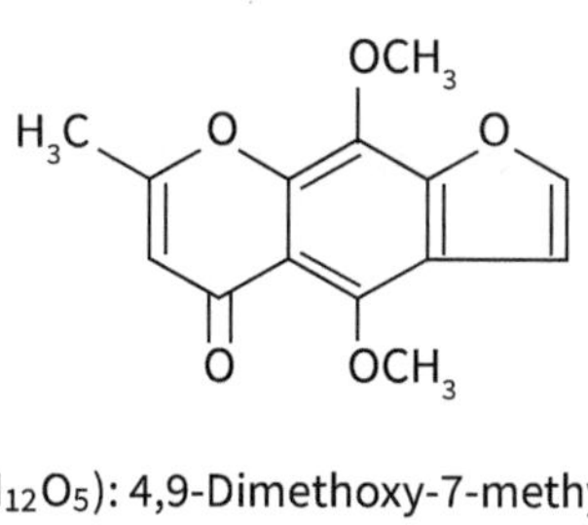

Figure 7.1 Kellin molecule ($C_{14}H_{12}O_5$): 4,9-Dimethoxy-7-methyl-5H-furo[3,2-g][1]benzopyran-5-one.

Lyotropic Liquid Crystals. Ingo Dierking and Antônio Martins Figueiredo Neto, Oxford University Press.
 DOI: 10.1093/9780191924927.003.0007

concentrations of molecules in the solution, small columns may exist. The increase in the number of molecules favours molecular packing in bigger columns and, as a consequence of column–column interactions, liquid crystalline structures are stabilized. This process, and the structural characteristic of the molecules, do not lead chromonics to show an equivalent to the Kraft temperature, observed in lyotropic phase diagrams with amphiphilic molecules [2].

7.2 Types of molecules in chromonics and molecule–molecule interaction

Figure 7.2 shows examples of molecules that form, in the presence of a solvent, chromonic mesophases. A common feature of them is the presence of aromatic rings and an almost flat atomic disposition. Interactions between neighbour molecules, arranged face-to-face, have their origin in the usual van der Waals atom–atom interactions; nevertheless $\pi - \pi$ interactions play an important role in the stabilization of the stacking structure [3]. Hunter and Sanders [3] proposed a model assuming the σ-framework and the π-electrons, separately. They showed that the net $\pi - \pi$ interaction comes from the $\pi - \sigma$ attractions, that overcome $\pi - \pi$ repulsions. These interactions stabilize the stacking structure, allowing also the slipping of molecules in the face-to-face arrangement. Due to the atomic (and charge) arrangement in the almost planar molecules, in most cases, a molecule and its first neighbours are not piled up exactly one over the other, but a little bit slipped in a direction that keeps the parallelism between them. This 'electrostatic accommodation' is followed by the molecules in the formation of the stacks. The local structure may be sketched as two negatively charged π-electron clouds separated by a positively charged σ-framework [4]. The energy involved in the molecule–molecule interaction, that stabilizes the stacks, is of the order of about 10 $k_B T$, where k_B and T are the Boltzmann constant and the absolute temperature.

When this type of molecule is in solution (usually in water), solvent molecules are expelled from the region between molecules that form stacks. At low molecular concentration, the mixture is isotropic (typical molar concentration $< 10^{-5} M$), with stacks randomly oriented. This situation drastically changes when the molecular concentration increases, with the stabilization of anisotropic mesophases.

7.3 Structures and phase diagrams

The most common structures encountered in chromonics are the nematic (labelled N) and the hexagonal phases (labelled M) [5]. Figure 7.3 shows a sketch of the aromatic molecular stacks in the isotropic (7.3a), nematic (7.3b), and hexagonal (7.3c) mesophases. The textures of these mesophases in a polarized optical microscope are shown in Figure 7.4. In some mixtures, the transition from the N to the M phase is of weak first order, and the evolution of the texture, as a function of

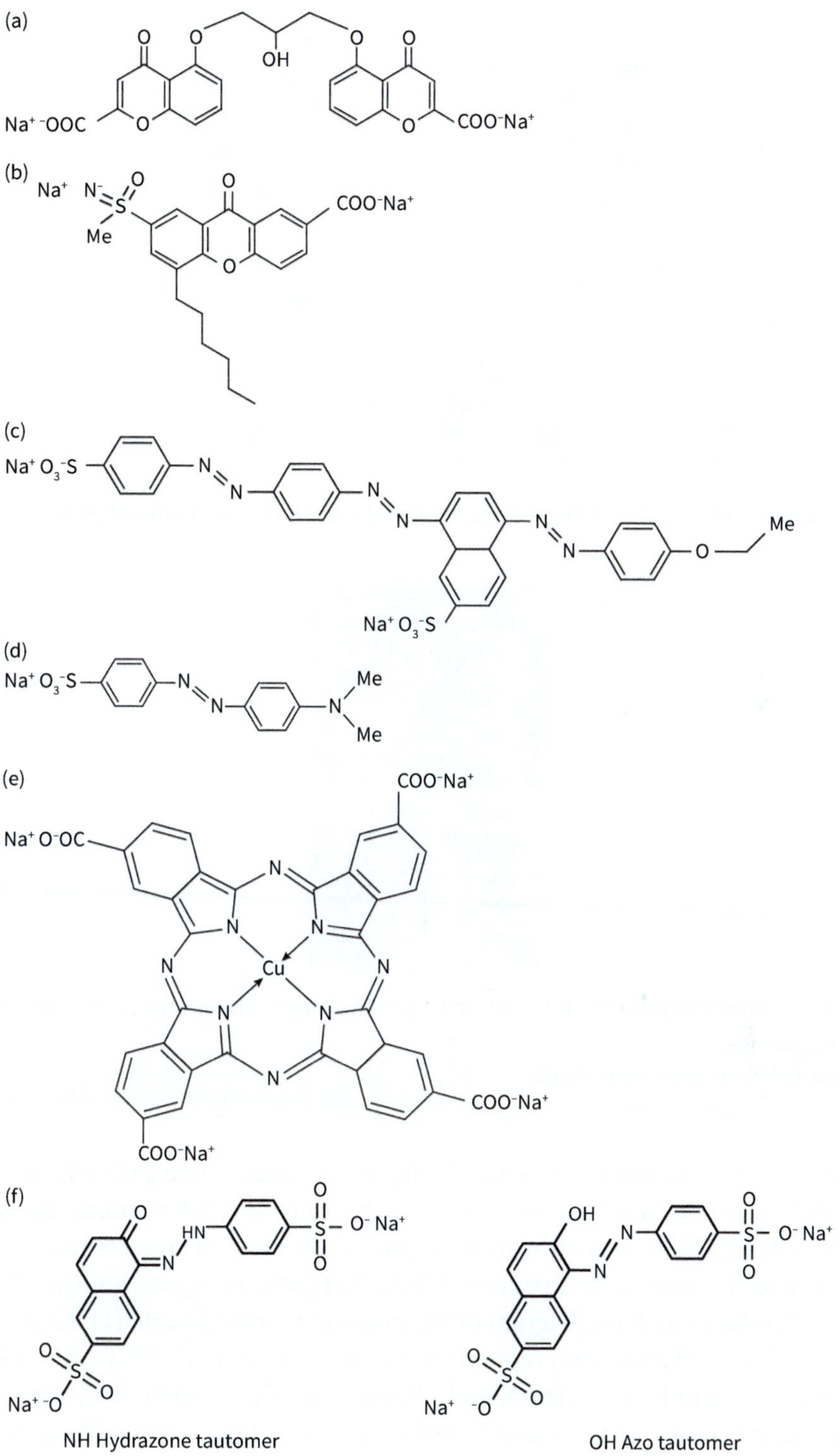

Figure 7.2 Examples of molecules that form chromonic liquid crystals [2]: (a) disodium cromoglycane; (b) RU31156; (c) Sirius supra brown RLL; (d) methyl orange; (e) copper-tetracarboxyphthalocyanine; (f) sunset yellow.

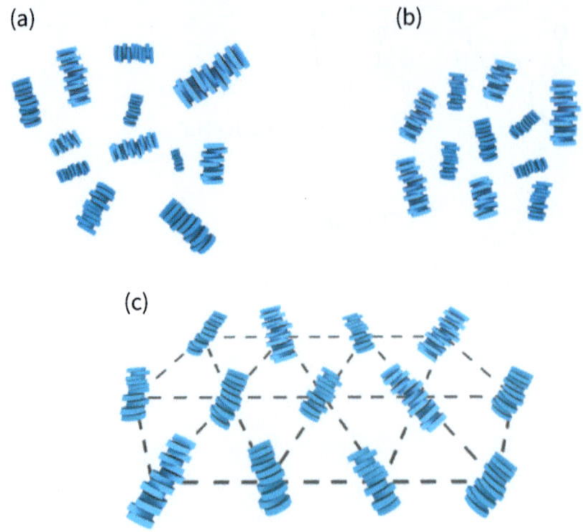

Figure 7.3 Sketches of the (a) isotropic Iso, (b) nematic N, and (c) hexagonal M phases.

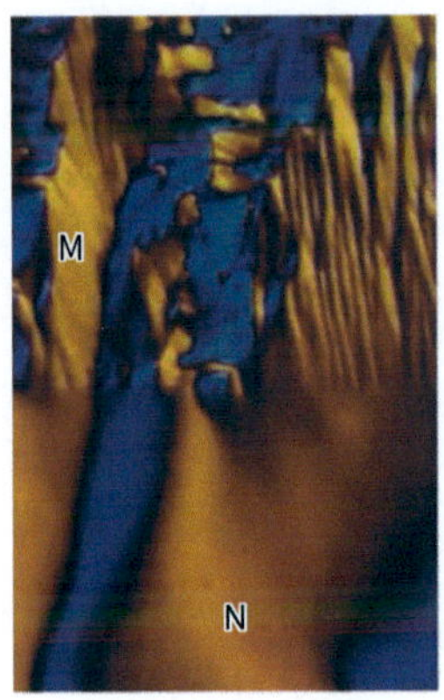

Figure 7.4 Optical textures of the N and M mesophases. N and M represent the nematic and hexagonal phases.
(Reproduced with permission from ref.[5]).

the temperature, was shown to be practically continuous. Some authors claim that this transition could be of second order, which, from symmetry considerations, is not allowed [6]. Tiddy and co-workers reported on other structures identified in aqueous mixtures of cyanine dye and CI Acid Red 266 (see Figure 7.5(a),(b)) [7].

Figure 7.6 shows a phase diagram of the disodium cromoglycate (DSCG)–water system [2]. The horizontal axis represents the concentration of DSCG in weight %. Coexistence regions between the N and M phases are clearly identified, supporting the characteristic first-order phase transition between these phases. As expected, by increasing the DSCG concentration, the solid phase is stabilized. At lowest concentrations, the isotropic and the nematic phases are stabilized.

Yamaguchi and co-workers reported on the structures encountered in sunset yellow/disodium cromoglycate/water mixtures [8]. The triangular phase diagram (at fixed (room) temperature) is shown in Figure 7.7. When these two types of

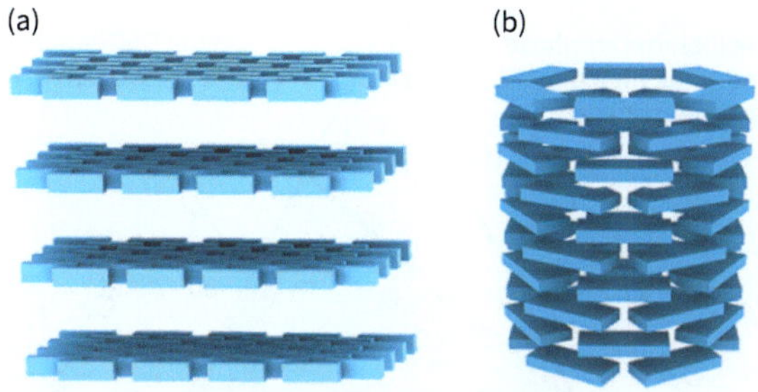

Figure 7.5 Structures proposed by Tiddy and co-workers to explain the diffraction data from the aqueous mixtures of (a) brickwork structure of the cyanine dye layer and (b) hollow tube structure of the CI Acid Red 266 aggregates.
(Reproduced with permission from ref.[2]).

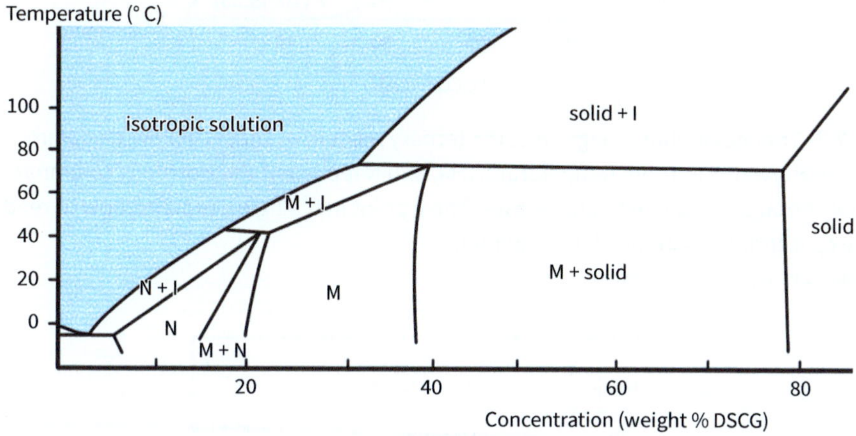

Figure 7.6 Phase diagram of the chromonic mixture of disodium cromoglycate/water. I, M, and N represent the isotropic, hexagonal, and nematic phases.
(Reproduced with permission from ref.[2]).

molecules are together, in water, independent columnar aggregates are formed, and very small regions of the nematic phase are present in the phase diagram. Mostly, biphasic regions are observed. A hexagonal phase was not clearly identified in the small-angle X-ray experiments, just the presence of columnar aggregates was proposed, and the authors suggested the label 'columnar phase' for the one observed in their experiments.

Molecules that form chromonic mesophases may be mixed with other materials to form other complex fluids. The phase diagram of the abietic acid/polyethylene glycol/water mixture is shown in Figure 7.8 [9]. The topology of the phase diagram, in this particular case, does not differ from what was described before. Other properties, however, may change.

The stability of the liquid crystalline mesophases in chromonics depends on the concentration of ions in the mixture, in particular H^{+} [10]. Acidic solutions lead to the formation of nonsoluble species in water, and molecular precipitation in the mixture.

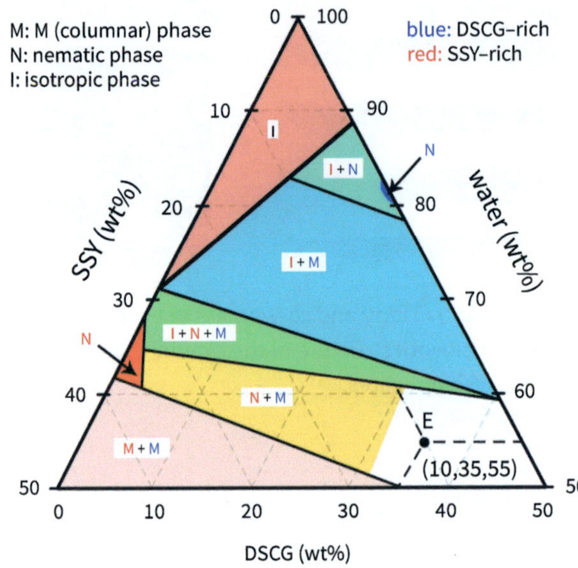

Figure 7.7 Triangular phase diagram of the ternary mixture of sunset yellow/disodium cromoglycate/water, at room temperature. I, M, and N represent the isotropic, columnar (probably hexagonal), and nematic phases. The E point in the figure explains how to read the relative concentrations in the phase diagram.
(Reproduced with permission from ref.[8]).

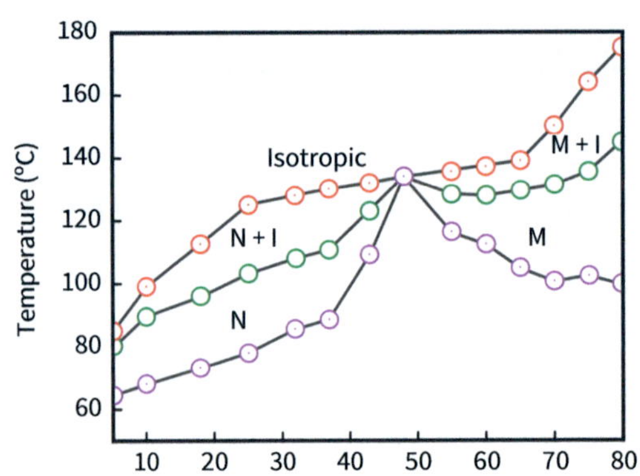

Figure 7.8 Phase diagram of the mixture abietic acid/polyethylene glycol/water. c represents the abietic acid molecule weight % concentration. I, M, and N represent the isotropic, hexagonal, and nematic phases.

Let us discuss the nanoscopic structure of the molecular aggregates present in chromonics and their structural arrangement. Figure 7.9 shows the phase diagram of the sunset yellow FCF (SSY)/water mixture [11]. The X-ray scattering intensity as a function of the modulus of the scattering wave vector ($q = (4\pi\, sin\theta)/\lambda$, where 2θ and λ are the scattering angle and X-ray wavelength) in isotropic and nematic mesophases is shown in Figure 7.10 [11]. In the isotropic phase, as a function of the increasing SSY molar concentration, the broad band located at about $q \sim 2\, nm^{-1}$,

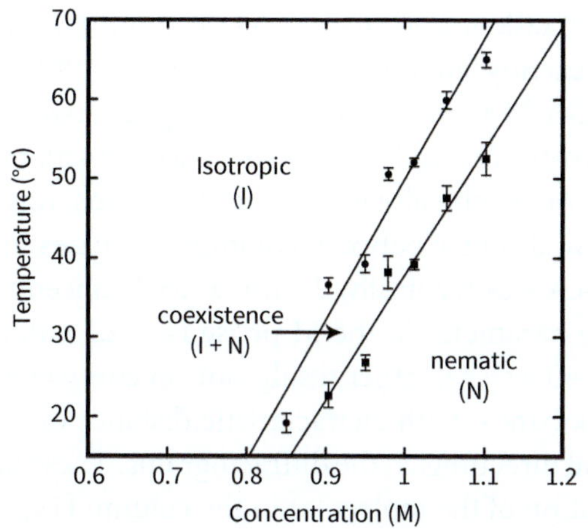

Figure 7.9 Phase diagram of the sunset yellow FCF (SSY)/water mixture. M represents the molar concentration of SSY. Solid lines are evaluated limits of the phase-coexistence region.

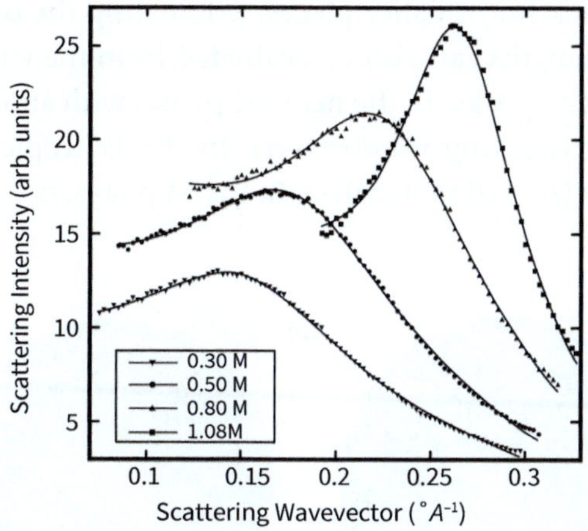

Figure 7.10 X-ray scattering intensity as a function of the scattering wave vector of isotropic and nematic mesophases. SSY/water mixture at 20 °C. Solid lines are Lorentzian functions fitted to the experimental results.
(Reproduced with permission from ref.[11]).

moves to higher values of q and becomes narrower, indicating higher correlation lengths. In the nematic phase, the band is narrower still, located around $q \sim 2.5\ nm^{-1}$. All the bands could be fitted to Lorentzian functions. A significant challenge to experimentalists is to orient chromonic mesophases to perform scattering and optical (e.g., birefringence) measurements. The strategy commonly used is to (partially) orient the columnar aggregates by flow in sample holders. As expected, the birefringence, that is related to the orientational order parameter in nematics, increases (in absolute values) as the molar concentration of SSY increases [11].

A comprehensive structural description of the N and M phases was reported by Joshi and co-workers [12]. The diffraction patterns of both phases, oriented by magnetic fields, were obtained. It is well known that orienting the chromonic nematic mesophase is a difficult task, unfortunately, details about this achievement were not given in the paper. Recently, attempts to mix ferrofluids in SSY/water mixtures to improve the magnetic-field alignment were not very successful. The diffractograms of the N and M phases are shown in Figures 7.11(a) and (b), respectively [12]. The distance between SSY molecules along the column axes remains practically constant (~ 0.33 *nm*). The mean distance between columns decreases in the N phase, as the SSY concentration increases (typically, 2.4 *nm* at SSY concentration of ~ 32 *wt*%). The hexagonal lattice parameter in the M phase is ~ 2.3 *nm*, at concentrations of SSY in the order of ~ 40 *wt*%. Another result, not yet completely understood, is the presence of four weak bands, with characteristic distance of ~ 0.56 *nm*, out of the vertical and horizontal directions in the diffractograms. These bands seem to be due to the oblique orientation of the molecules in the column [13].

The growth of the columnar aggregates is well described by a model where the molecules tend to stack face-to-face to reduce the contact of particular areas with the solvent, but this process is limited for entropic reasons. This balance leads to the breaking of a long stack into smaller pieces, originating the different mesophases [14]. The correlation lengths can also be evaluated from the width at half height of the diffraction bands and peaks. In the nematic phase, with about 15 SSY molecules forming the columns (packing number—*n*). In the isotropic phase the packing number is still smaller (*n* < 10 molecules). In the M phase, *n* ~ 20 [12]. The growth

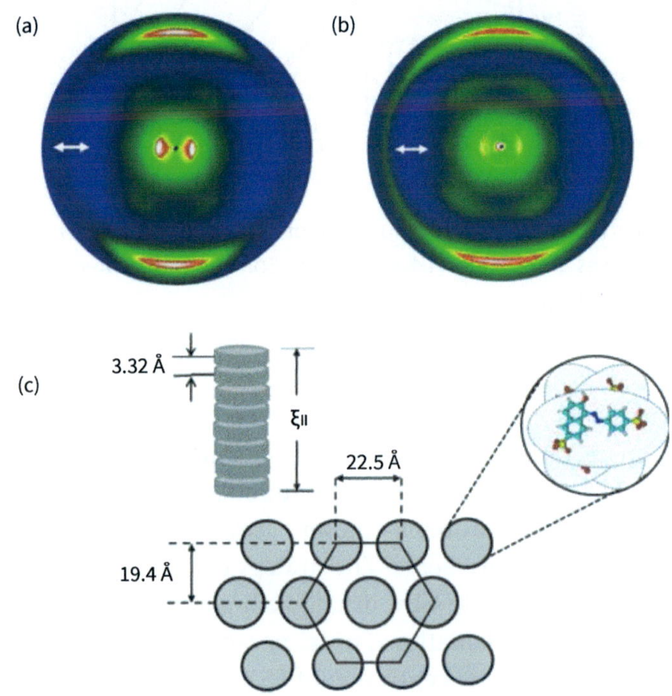

Figure 7.11 Diffractograms of the (a) N and (b) M magnetically (white double arrow) oriented mesophases. SSY/water mixture. (c) Typical dimensions obtained from the diffraction bands and peaks, and suggested columnar structure and arrangement.
(Reproduced with permission from ref.[12]).

of the stacks as a function of the molecular concentration seems to be isodesmic [15], i.e., the stack grows continuously, without a jump in n, at the phase transition.

7.4 Elastic properties

Different from the usual thermotropic LCs and lyotropic mixtures with amphiphilic molecules, chromonics have the characteristics of a strongly modified shape anisotropy of their basic units (the columns) as a function of molecular concentration. As discussed in the previous sections, the noncovalent attraction between molecules promotes the face-to-face stacking of them, forming columns whose length varies from that of a single molecule to about 20 piled molecules. The stacking process depends on the relative concentration of the solute and solvent, and the temperature. This aspect allows chromonics to show elastic properties particularly interesting and tunable, depending on the temperature and stacking degree.

The length of a stack (L_c) can be evaluated by using the theoretical approach proposed by van der Schoot and Cates [16], in the framework of wormlike micelles interacting via excluded volume. This parameter is essential in the interpretation of elastic constant measurements [17]. Figure 7.12 shows the ratio between the Frank elastic constants [18] of the sunset yellow/water mixture (nematic phase), for different temperatures and relative concentrations [17]. The values of the elastic constants reported are: $K_{11} = (4.3 \pm 0.4)\ pN$; $K_{22} = (0.70 \pm 0.07)\ pN$; $K_{33} = (6.1 \pm 0.6)\ pN$, at the concentration of SSY of 29%. Different from the usual thermotropic liquid crystals, where the three elastic constants are of the same order of magnitude [19,20], K_{22} is about one order of magnitude smaller than the other elastic constants. At low temperatures, K_{11} and the ratios K_{11}/K_{22}, K_{11}/K_{33} increase as the solute concentration increases. This behaviour was attributed to the increase of the stack length, as the solute concentration increases (Obs.: the labels K_i and K_{ii} are used indistinctly).

Similar results were encountered in the nematic phase of the mixture disodium cromoglycate (DSCG)/water [21]. Figure 7.13 depicts the Frank elastic constants at this mixture, at different temperatures and relative concentrations of the solute (c). The order of magnitude of the bend elastic constant in this system is the same as obtained for amphiphilic lyotropic liquid crystals, like mixtures of potassium laurate/decanol/water and sodium decylsulfate/decanol/water, or in the nematic calamitic phase [22]. The viscosities (η_{splay}, η_{twist}, η_{bend}) of the DSCG/water mixture, in the nematic phase, were measured by using Dynamic Light Scattering [21]. At low temperatures, the higher the solute concentration, the larger the viscosity. At a fixed solute concentration, η decreases as the temperature increases. The order of magnitude of η_{bend} is about 4 times smaller than that of both η_{splay} and η_{twist}. Just to give values for these parameters, at ~ $294.5K$ and $c = 18\ wt\%$, $\eta_{splay,twist}$ ~ $300\ kg\ m^{-1}s^{-1}$ and η_{bend} ~ $0.03\ kg\ m^{-1}s^{-1}$. The order of magnitude of the bend viscosity η_{bend} is about the same as encountered in thermotropics. The higher values of $\eta_{splay,twist}$ are due to the presence of long stacks (length much larger than its diameter) in the solution.

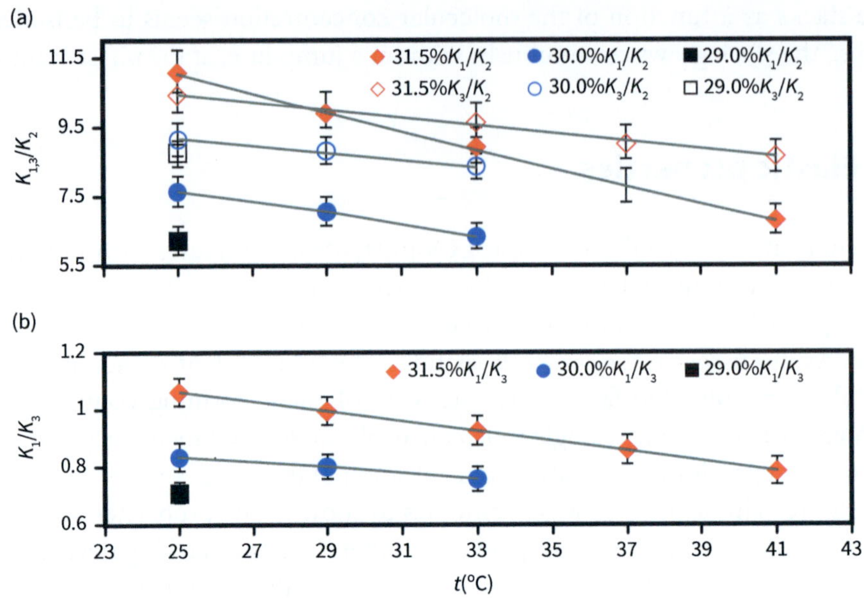

Figure 7.12 Mixture sunset yellow/water. Ratio between elastic constants for different relative concentrations and temperature. (a) $K_{1,3}/K_2$; (b) K_1/K_3.

An attempt to theoretically describe the elastic constants' behaviour was undertaken by Romani and co-workers [23]. The model assumes the molecules as discs (diameter D and thickness $\mathcal{L}_d$), with $\mathcal{L}_d/D = 0.31$, that interact face-to-face to form polydisperse columnar aggregates. An additional energetic cost for splay deformation is introduced in the model, originating from the coupling between the areal density and the local director field. Monte Carlo simulations considering flexible and polydisperse stacks allowed the prediction of some of the behaviours of the ratios between elastic constants measured in chromonics.

The smaller value of the twist elastic constant, with respect to those of the splay and bend constants, found in chromonics, seems not to be a characteristic only of this category of liquid crystals. Recently, Dietrich and co-workers [24] measured the three elastic constants of the lyotropic mixture N, N-dimethyl-n-ethylhexadecylammonium bromide (CDEAB)/decan-1-ol/water, in the discotic nematic phase ($T = 298\ K$). Differing from chromonics, this mixture, in the nematic phase, has micelles as basic units. Light-scattering experiments were employed to measure the elastic constants: $K_{11} = (1.54 \pm 0.08)\ pN$; $K_{22} = (0.300 \pm 0.018)\ pN$; $K_{33}=(20.8 \pm 1.3)\ pN$. Here, K_{22} is one order of magnitude smaller than K_{11} and two orders of magnitude smaller than K_{33}. In both lyotropic chromonic and lyotropic micellar systems, the basic units (stacks and micelles, respectively) show a certain degree of flexibility. This aspect, not present in single-molecule thermotropics, was suggested to be the responsible factor for the anisotropy of the elastic constants experimentally encountered in lyotropics.

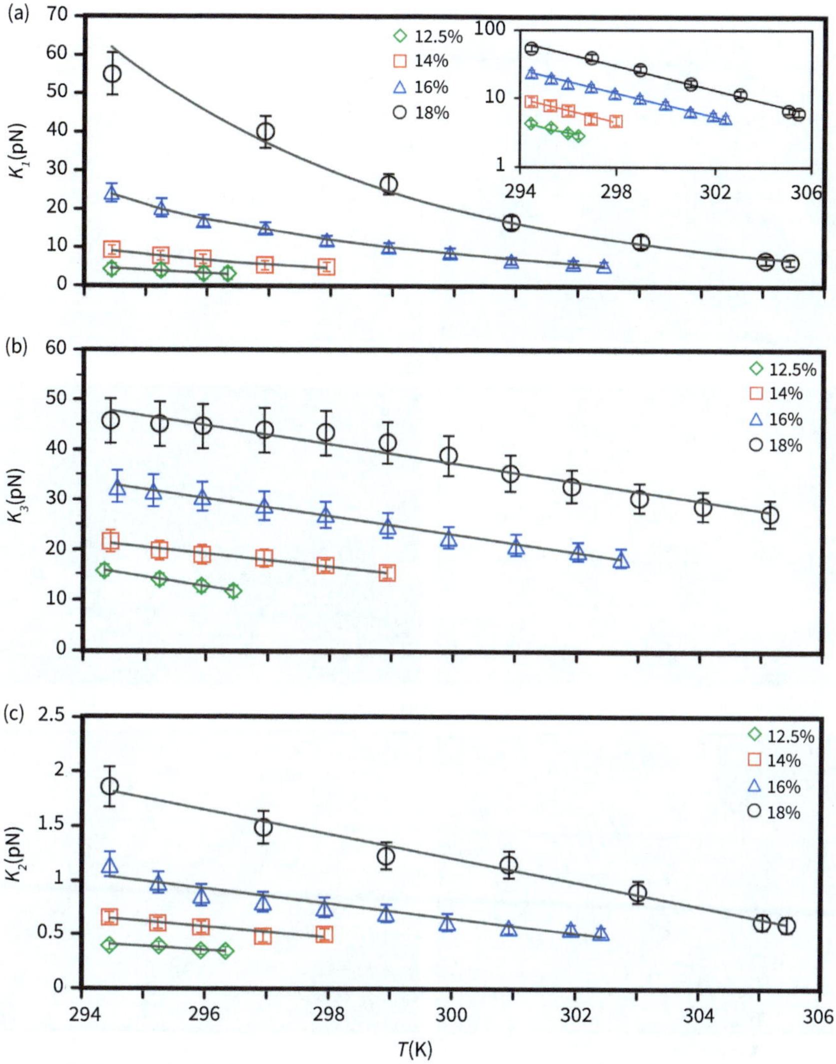

Figure 7.13 Frank elastic constants of the mixture disodium cromoglycate/water, in the nematic phase, for different relative concentrations (in weight %), as a function of the temperature. (a) K_1; (b) K_3; (c) K_2.

7.5 Textures and defects

The observation of the textures of chromonic thin films, under polarized optical microscopy, revels not only spectacular images from the aesthetic point of view, but also allows the evaluation of the stack's arrangement in the different mesophases. Figure 7.14 shows typical textures of the nematic (N) and hexagonal (M) mesophases, as well as biphasic domains, where the isotropic phase is present, for a mixture of disodium cromoglycate (DSCG)/water [5]. Textures of the N and M mesophases of the mixture SSY/water are shown in Figure 7.15(a),(b) [12].

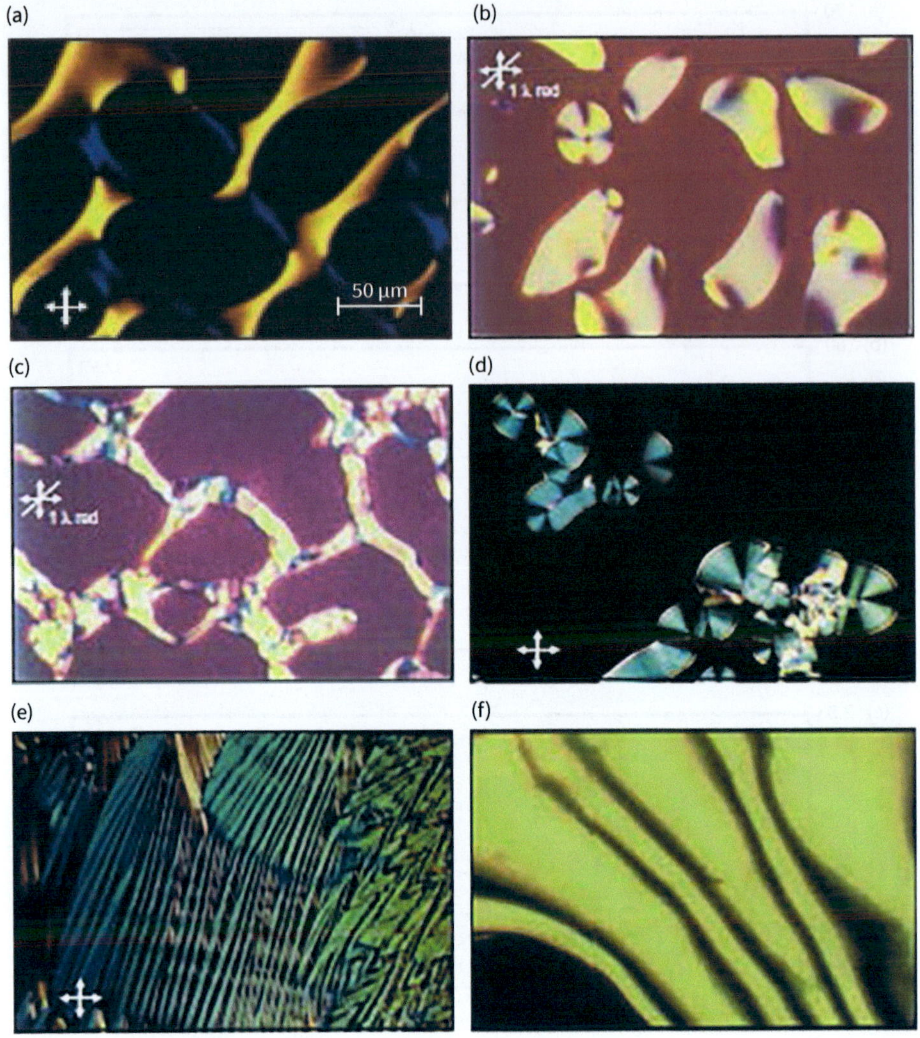

Figure 7.14 Typical textures of chromonic mesophases. Mixture of disodium cromoglycate (DSCG)/water. Films under crossed polarizers (directions indicated by white arrows). (a), (b) nematic (N) and isotropic (I) biphasic domains; (c), (d) hexagonal (M) and I biphasic domain; (e) N phase; (f) M phase.

(Reproduced with permission from ref.[5]).

In a seminal paper, Ericksen discussed spatially configurations of the director ($\vec{n}$) in uniaxial nematics and calculated, assuming uniform gradients for $\vec{n}$, inequalities between elastic constants [25]: $0 < K_{24} < 2K_{22}$ and $K_{24} < 2K_{11}$, where K_{24} is the saddle-splay surface elastic constant. As discussed in the previous section, K_{22} in chromonics is one order of magnitude smaller than those of the splay and bend elastic constants. Davidson and co-workers investigated the SSY/water mixture, in the nematic phase, confined in a hollow cylinder with degenerate planar boundary conditions [26]. In their experiments, hedgehog defects were observed, characteristic of

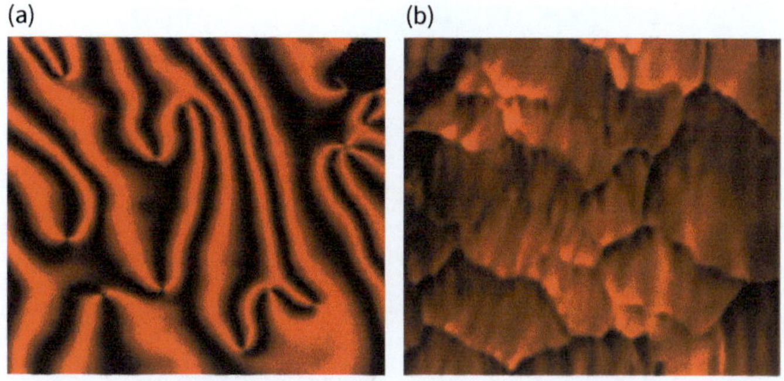

Figure 7.15 Typical textures of chromonic mesophases. Mixture of sunset yellow (SSY)/water. Films under crossed polarizers. (a) N phase; (b) M phase.
(Reproduced with permission from ref.[12]).

an escaped-twist configuration. The ratio K_{24}/K_{22} measured was about 50, clearly violating Ericksen's inequality.

Due to the size and shape anisotropy of the basic units of chromonics (molecular stacks), they represent an excellent system to observe defects, e.g., disclinations, in the topological arrangement of the stacks, in the nematic phase. This characteristic of the stacks allows a detailed analysis of these defects, in particular their cores, by using the LC-PolScope and Cryo-TEM experimental setups [27]. In particular, the LC-PolScope maps the optical phase retardance and the projection of the director in the sample-surface plane. Findings about the dependence of the director and scalar order parameter with azimuthal and radial coordinates in chromonic nematics are expected to be present also in thermotropics, but it has not been experimentally encountered to date.

Recently, Zhang and co-workers [28] reported on an interesting experiment, where the chromonic mixture of DSCG/water mixture in the nematic phase is injected into a rectangular microfluidic channel of length 50 mm, width 15 mm, and thickness (6.5 ± 1) μm. The flow range employed was from 0.07 to 25 $\mu L/min$. Pure-twist disclination loops were generated, with size and dynamics depending on the shear rate of the liquid crystal.

7.6 Chromonics doped with chiral molecules

The doping of amphiphilic–lyotropic nematics with chiral agents is a common procedure in the field of liquid crystals (see Chapter 6). Cholesteric mesophases are stabilized for different types of studies in both the bulk and in the micellar layers at interfaces. Chiral molecules employed in this task may be incorporated in the micelles (e.g., chiral amphiphiles) or located in the liquid medium between micelles. In the case of chromonics, doped with chiral molecules, the number of studies published until now is small.

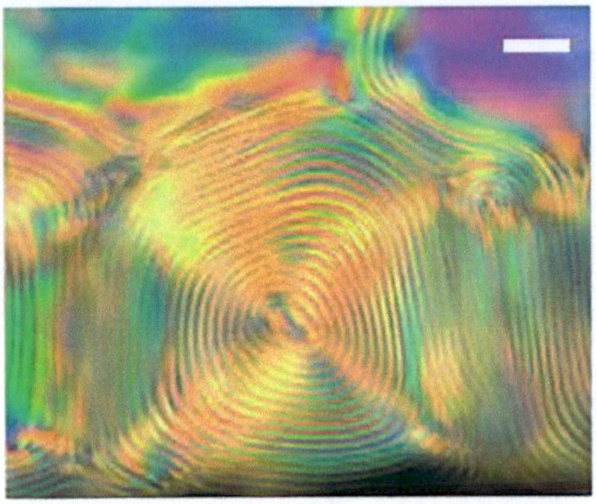

Figure 7.16 Textures of the mixture DSCG/l-alanine/water, in cross-polarized optical microscope. The concentration of l-alanine is 7 wt%. Scale bar 50 mm. (Reproduced with permission from ref.[29]).

In 2015, Peng and Lavrentovich reported a study where aqueous solutions of disodium cromoglycate (DSCG) and polyethyleneglycol (employed to stabilize tactoids) were doped with L-alanine, limonene or brucine. Due to the elastic properties of the nematic phase, transformed to a cholesteric by the chiral molecules, an amplification of the chirality was observed in the tactoids present in the mixture [29]. The optical activity was amplified by orders of magnitude with respect to isotropic media with the same amount of chiral dopants (Figure 7.16). The mechanism responsible for this phenomenon is still under discussion.

The pitch (P) of aqueous mixtures of DSCG, doped with L-alanine or L-proline was measured as a function of the dopant concentration (c) [30]. A linear dependence of P^{-1} with c was observed at low dopant concentrations, as theoretically expected. The order of magnitude of the helical superstructure corresponding to $c \sim 5\ wt\%$ of L-alanine is $P \sim 20\ \mu m$.

The effect of chiral dopants in chromonics, confined in cylindrical geometry, was reported in [31]. Due to the characteristic values and rapport between elastic constants in nematic lyotropic chromonics (without any chiral dopant), when confined in cylindrical geometry, a double-twist chiral director configuration appears. Adding a chiral dopant (above a critical concentration), this double-twisted configuration transforms to homochiral, revealing striking properties of these systems.

References

[1] J. Lydon (2007). A personal history of the early days of chromonics. *Liquid Crystals Today*, 16, 13–27.

[2] J. E. Lydon (1998). Chromonic liquid crystal phases. *Curr. Opin. Colloid Interface Sci.*, 3, 458–466.

[3] C. A. Hunter and J. K. M. Sanders (1990). The nature of π–π interactions. *J. Am. Chem. Soc.*, 112, 5525–5534.

[4] J. E. Lydon (1998). In: D. Demus, J. Goodby, G. W. Gray, H. W. Speiss, and V. Vill (eds), *Handbook of Liquid Crystals*, vol 2B,. Weinheim: Wiley VCH.

[5] J. Lyndon (2011). Chromonic liquid crystalline phases. *Liquid Crystals*, 38,1663–1681.

[6] J. C. Tolédano and P. Tolédano (1987). The Landau Theory of Phase Transitions, *World Scientific Lecture Notes in Physics*, vol 3, Singapore: World Scientific Publishing.

[7] G. J. T. Tiddy, D. L. Mateer, A. P. Ormerod, W. J. Harrison, and D. J. Edwards (1995). Highly ordered aggregates in dilute dye-water systems. *Langmuir*, 11, 390–393.

[8] A. Yamaguchi, G. P. Smith, Y. Yi, C. Xu, S. Biffi, F. Serra, T. Bellini, C. Zhu, and N. A. Clark (2016). Phases and structures of sunset yellow and disodium cromoglycate mixtures in water. *Phys. Rev. E*, 93, 012704.

[9] T. N. Govindaiah and H. R. Sreepad (2015). Chromonic liquid crystalline phase transition involving a biphasic region of nematic and hexagonal phases. *Phase Transitions*, 88, 183–191.

[10] B. Zhang and H. S. Kitzerow (2016). Influence of proton and salt concentration on the chromonic liquid crystal phase diagram of disodium cromoglycate solutions: Prospects and limitations of a host for DNA nanostructures. *J. Phys. Chem. B*, 120, 3250–3256.

[11] V. R. Horowitz, L. A. Janowitz, A. L. Modic, P. A. Heiney, and P. J. Collings (2005). Aggregation behavior and chromonic liquid crystal properties of an anionic monoazo dye. *Phys. Rev. E*, 72, 041710.

[12] L. Joshi, S. W. Kang, D. M. Agra-Kooijman, and S. Kumar (2009). Concentration, temperature, and pH dependence of sunset-yellow aggregates in aqueous solutions: An X-ray investigation. *Phys. Rev. E*, 80, 041703.

[13] H. S. Park, S. W. Kang, L. Tortora, Y. Nastishin, D. Finotello, S. Kumar, and O. D. Lavrentovich (2008). Self-assembly of lyotropic chromonic liquid crystal sunset yellow and effects of ionic additives. *J. Phys. Chem. B*, 112,16307–16319.

[14] W. M. Gelbart and A. Ben-Shaul (1996). The 'new' science of 'complex fluids'. *J. Phys. Chem.*, 100, 13169–13189.

[15] M. M. J. Smulders, M. M. L. Nieuwenhuizen, T. F. A. de Greef, P. van der Schoot, A. P. H. J. Schenning, and E. W. Meijer (2010). How to distinguish isodesmic from cooperative supramolecular polymerisation. *Chem. Eur. J.*, 16, 362–367.

[16] P. van der Schoot and M. E. Cates (1994). The isotropic-to-nematic transition in semi-flexible micellar solutions. *Europhys. Lett.*, 25, 515–520.

[17] S. Zhou, Y. A. Nastishin, M. M. Omelchenko, L. Tortora, V. G. Nazarenko, O. P. Boiko, T. Ostapenko, T. Hu, C. C. Almasan, S. N. Sprunt, J. T. Gleeson, and O. D. Lavrentovich (2012). Elasticity of lyotropic chromonic liquid crystals probed by director reorientation in a magnetic field. *Phys. Rev. Lett.*, 109, 037801.

[18] P. Sheng (1976). In: E. B. Priestley, P. J. Wojtowicz, and P. Sheng (eds), *Introduction to Liquid Crystals*, New York: Plenum Press.

[19] A. Bogi and S. Faetti (2001). Elastic, dielectric and optical constants of 4′-pentyl-4-cyanobiphenyl. *Liquid Crystals*, 28, 729–739.

[20] W. Iglesias, N. L. Abbott, E. K. Mann, and A. Jákli (2012). Improving liquid-crystal-based biosensing in aqueous phases. *Appl. Mater. Interfaces*, 4, 6884–6890.

[21] S. Zhou, K. Neupane, Y. A. Nastishin, A. R. Baldwin, S. V. Shiyanovskii, O. D. Lavrentovich, and S. Sprunt (2014). Elasticity, viscosity, and orientational fluctuations of a lyotropic chromonic nematic liquid crystal disodium cromoglycate. *Soft Matter*, 10, 6571–6581.

[22] T. Kroin, A. J. Palangana, and A. M. Figueiredo Neto (1989). Determination of the bend elastic constant and the anisotropy of diamagnetic susceptibility of lyotropic nematic calamitic liquid crystals. *Phys. Rev. A*, 39, 5373–5377.

[23] E. Romani, A. Ferrarini, and C. De Michele (2018). Elastic constants of chromonic liquid crystals. *Macromolecules*, 51, 5409–5419.

[24] C. F. Dietrich, P. J. Collings, T. Sottmann, P. Rudquist, and F. Giesselmann (2020). Extremely small twist elastic constants in lyotropic nematic liquid crystals. *PNAS*, 117, 27238–27244.

[25] J. L. Ericksen (1966). Inequalities in liquid crystal theory. *The Physics of Fluids*, 9, 1205–1207.

[26] Z. S. Davidson, L. Kang, J. Jeong, T. Still, P. J. Collings, T. C. Lubensky, and A. G. Yodh (2015). Chiral structures and defects of lyotropic chromonic liquid crystals induced by saddle-splay elasticity. *Phys. Rev. E*, 91, 050501(R).

[27] S. Zhou, S. V. Shiyanovskii, H.-S. Park, and O. D. Lavrentovich (2017). Fine structure of the topological defect cores studied for disclinations in lyotropic chromonic liquid crystals. *Nature Communications*, 8, 14974.

[28] Q. Zhang, R, Zhang, B. Ge, Z. Yaqoob, P. T. C. So, and I. Bischofberger (2021). Structures and topological defects in pressure-driven lyotropic chromonic liquid crystals. *PNAS*, 118, e2108361118.

[29] C. Peng and O. D. Lavrentovich (2015). Chirality amplification and detection by tactoids of lyotropic chromonic liquid crystals. *Soft Matter*, 11, 7257–7264.

[30] T. Ogollaa, S. B. Nasheda, and P. J. Collings (2017). Pitch measurements in chiral lyotropic chromonic liquid crystals. *Liquid Crystals*, 44, 1968–1978.

[31] J. Eun, S. Kim, and J. Jeong (2019). Effects of chiral dopants on double-twist configurations of lyotropic chromonic liquid crystals in a cylindrical cavity. *Phys. Rev. E*, 100, 012702.

8
Inorganic lyotropic liquid crystals

Given the more recent revival of lyotropic phases formed by anisotropic nanoparticles, it might sound surprising, but the field of inorganic lyotropic liquid crystals goes back more than a century in time. It was in 1902 when Majorana [1] discovered the possibility of inducing birefringence in an aqueous colloidal suspension of iron oxides when applying a magnetic field, the equivalent of the Cotton–Mouton effect [2] in molecular liquids.

In common with thermotropic liquid crystals, inorganic or mineral lyotropic liquid crystals are also based on anisotropic building blocks. While in the former case these are rod- or disc-like molecules, in the latter case these are corresponding natural, mineral needles or plates of nanometre size. The description of ordering in the thermotropic case is accomplished via the well-known Maier–Saupe theory [3a,b,c]. This is based on induced dipole–induced dipole interactions of elongated molecules. It predicts a temperature dependence of the scalar order parameter S changing discontinuously at the isotropic to nematic transition from $S=0$ to $S=0.4$ and then monotonically increasing to approximately $S\sim0.7–0.8$ for decreasing temperature.

In the case of lyotropic mineral liquid crystals the ordering mechanism was first described by Onsager [4] who started from an opposite point of view. All interactions were purely repulsive, attractive van der Waals interactions were neglected, and the sole assumption was that two particles cannot occupy the same space at the same time. Repulsive interactions were assumed to be mainly steric in nature, thus being based on excluded volume effects, but could also be electrostatic. This electrostatic repulsion was already included by Onsager as an effective increase in particle size, before being brought into more elaborate form much later by Stroobants [5] and Vroege [6].

In the Onsager theory the aspect ratio between diameter D and length L, thus D/L, plays a central role in the description of ordering. The formation of phases, i.e., spontaneous ordering of anisotropic colloidal particles, is described by the minimization of the free energy $F=U-TS$ with the internal energy U being $U=0$ for purely steric repulsion, T being the temperature, and S the entropy. In general, there are two competing effects. A decrease of the excluded volume implies an increase of the free volume, and thus also an increase of the entropy of translation and a reduction of the entropy of mixing. When the colloidal particle concentration is small the mixing entropy dominates the phase behaviour, and the isotropic phase is observed. Once a critical concentration is reached, the formation of the nematic state is observed when entering the two-phase region. Further increase of the concentration shows a monotonic increase of the liquid crystalline volume

Lyotropic Liquid Crystals. Ingo Dierking and Antônio Martins Figueiredo Neto, Oxford University Press.
 DOI: 10.1093/9780191924927.003.0008

fraction until the whole sample is liquid crystalline on leaving the two-phase region. The nematic order parameter increases at rather high values in the nematic phase from about S=0.8 to approximately S~0.95, generally larger than observed for thermotropic systems. From Onsager theory it was further found that the isotropic phase is observed for volume fractions $\psi_{ISO}<3.3D/L$ while the two-phase region terminates for $\psi_N>4.5D/L$.

When discussing inorganic or mineral liquid crystals we will realize that there is an overlap with one of the later chapters on lyotropic liquid crystals from nanorods, nanotubes, and nanosheets. Both are closely related and the borderline between these systems is flowing and diffuse.

8.1 Rod-like mineral materials

The first of the inorganic liquid crystalline materials to be discovered, and one of the most investigated ones is surely vanadium pentoxide, V_2O_5. First studies go back more than a century when Freundlich realized that a solution containing vanadium pentoxide becomes birefringent on stirring [7]. This was confirmed by Zocher [8] who investigated the flow-induced birefringent in much more detail.

8.1.1 Vanadium pentoxide, V_2O_5

Vanadium pentoxide is a yellowish-orange powder (Figure 8.1(a)), which at low concentrations produces a brownish, yet transparent solution with water. The crystal structure of V_2O_5 is shown in Figure 8.1(b) [9]. At large magnification in scanning electron microscopy (SEM, Figure 8.1(c)) [10], it becomes clear that vanadium pentoxide forms needle-like crystallites of strongly elongated, rod-like form which were shown to be nano-ribbons [11]. Only a few years after the discovery of flow birefringence and its quantitative investigation [12], Zocher also realized that V_2O_5 exhibits a 'voluntary structural formation in sols' [13], as he calls it, 'voluntary' meaning without any application of external forces. He thus reports the spontaneous formation of an ordered state and therefore liquid crystalline behaviour for the V_2O_5 colloids in water; a lyotropic liquid crystal.

It has repeatedly been reported that a freshly prepared sol from vanadium pentoxide does not exhibit any birefringence and that only after considerable time, days or sometimes even weeks, birefringent domains appear in the form of tactoids. This is depicted in Figure 8.1(d), a figure reproduced from the publication of Zocher [13]. Tactoids are domains with orientationally ordered rod-like particles whose shape is significantly different from the commonly observed spherical domains in thermotropic nematics growing within the sea of the isotropic phase. They are not only observed for V_2O_5 but for a whole range of inorganic materials forming lyotropic liquid crystals, and also for chromonic LCs and other lyotropics. Tactoids exhibit two pointed tips in the direction of the long axis of the particles and their structure was

Figure 8.1 (a) Vanadium pentoxide is probably the best studied of all the mineral liquid crystals. (b) Crystal structure of V_2O_5 with the large black spheres illustrating the vanadium atoms and the smaller orange ones the oxygen [9]. (c) The needle-like structure of vanadium pentoxide is imaged via SEM [10]. (d) The growth of liquid crystalline, nematic tactoids from the isotropic phase [13].

(Figure (b) reproduced with permission from ref.[9], Figure (c) reproduced with permission from ref.[10], Figure (d) reproduced with permission from ref.[13]).

discussed in detail by Kaznacheev et al. [14, 15] as well as theoretically by Prinsen and van der Schoot [16]. The structure of tactoids can additionally be influenced by applied electric [17] and magnetic [18] fields, which often results in a rounding of the pointed tips. Furthermore, the whole lyotropic nematic phase of V_2O_5 suspensions in water could be re-oriented by magnetic fields [19].

The shape of the V_2O_5 crystallites or ribbons and the lyotropic phase formation is related to the conditions during the synthesis, which was studied in detail by Pelletier et al. [20]. It is related to the aging of freshly prepared sols, which in turn depends on temperature, time, electrolyte addition, and of course the concentration of vanadium pentoxide in the fluid. The aging process, i.e., the formation of the nematic phase from the sol, increases for higher V_2O_5 concentrations, elevated temperatures, and the addition of electrolytes. The width of the ribbons is not changed significantly and is approximately constant at about 10 nm while the length increases from tens of nanometres into the micrometre regime, thus approaching the realm of nanowires. This is accompanied by a sol-gel transition [21–23] and it was shown that attractive interactions increased the tendency of nematic order in the gel [24]. Further, it could be demonstrated by nuclear magnetic resonance (NMR) that at sufficiently large concentrations of V_2O_5 ribbons the gel underwent a transition into a biaxial nematic gel phase [25].

The phase behaviour of the vanadium pentoxide lyotropic liquid crystal is thus given as depicted in Figure 8.2 [23]. At low volume concentrations of V_2O_5 one observes the isotropic liquid suspension which then starts to exhibit a two-phase

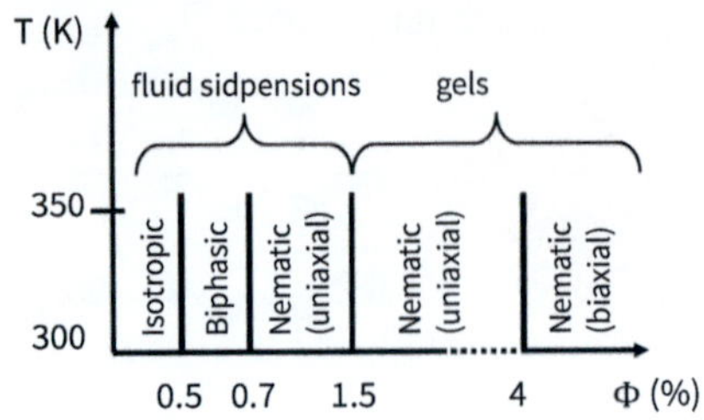

Figure 8.2 Phase diagram of vanadium pentoxide as a function of concentration [23].

behaviour between isotropic and nematic between volume concentrations of about 0.5–0.7. For increasing concentration, a uniaxial nematic sol is observed until the sol-gel transition occurs at a concentration of approximately $\Phi \approx 1.5$ with the formation of a uniaxial nematic gel, which becomes biaxial at much larger concentrations of about $\Phi \approx 4$. An early review especially on the synthesis, structure, and properties of V_2O_5 gels was provided by Livage [26] in 1991.

8.1.2 Aluminium oxyhydroxide, AlOOH

Our next example for an inorganic lyotropic liquid crystal is aluminium oxyhydroxide, AlOOH. The crystalline structure of aluminium oxyhydroxide is shown in Figure 8.3(a), consisting of octahedral layers that are connected via hydrogen bonds.

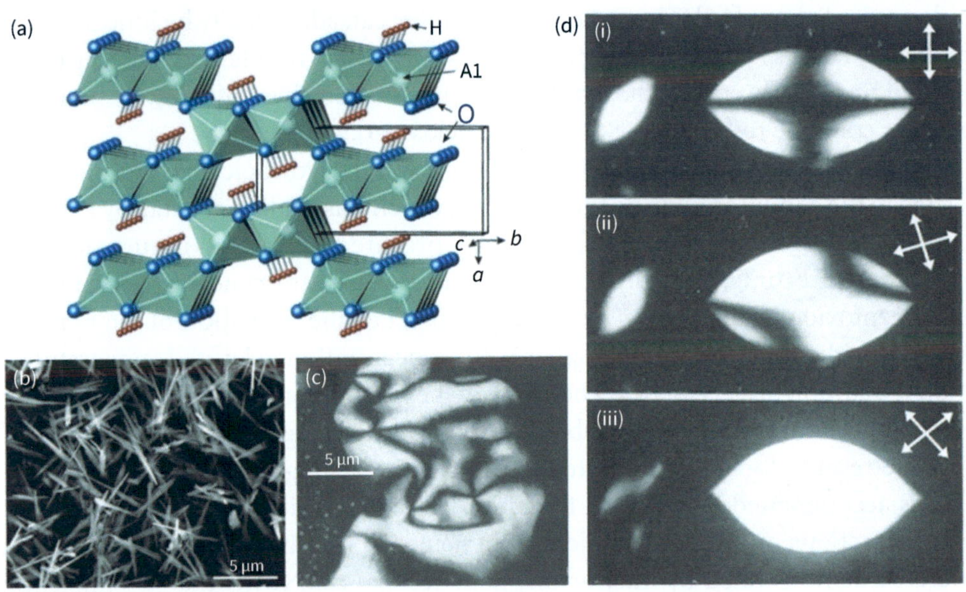

Figure 8.3 (a) Crystalline structure of AlOOH [28]. (b) SEM micrograph of the needle-like structure of AlOOH [29]. (c) Schlierentexture of the nematic phase of aluminium oxyhydroxide; and (d) AlOOH tactoids under different directions of the crossed polarizers, indicated as arrows [27].

(Figure (a) reproduced with permission from ref.[28], Figure (c) and (d) reproduced with permission from ref. [27]).

In the SEM image of the material (Figure 8.3(b)) one can again see needle-like, elongated crystallite aggregates, although not quite as nicely pronounced as the ones discussed above. The AlOOH sol was first synthesized by Zocher and Török [27] via dissolution of aluminium foil in acetic acid, under the influence of adding a small amount of mercury acetate. The solution was then boiled and cooled down, when it became opaque and birefringent. Addition of water then resulted in a lyotropic nematic AlOOH sol with a typical Schlierentexture as depicted in Figure 8.3(c). This Schlierentexture was formed by the merging of tactoids, which are shown in Figure 8.3(d) with the arrows indicating crossed polarizers. Rotating the polarizers through 45 degrees, as depicted in (i)–(iii) in the figure, produces textures which are in accordance with the expected director pattern within the tactoid.

8.1.3 Lithium molybdenoselenite, $Li_2Mo_6Se_6$

There is another whole class of materials, of which lithium molybdenoselenite, $Li_2Mo_6Se_6$, is a member, which also exhibits lyotropic nematic behaviour. Besides lithium, M=Li, the metal can also be other alkali metals such as M=Na, K, Rb, Cs, but also In, Ag, or Cu. Instead of X=Se also X=Te can be used. Of this general class of materials with general formula $M_2Mo_6X_6$ several members can be dissolved in highly polar liquids, but sol and gel phases are not stable. It is only lithium molybdenoselenite which forms long nanowires that exhibit a stable lyotropic phase [30,31].

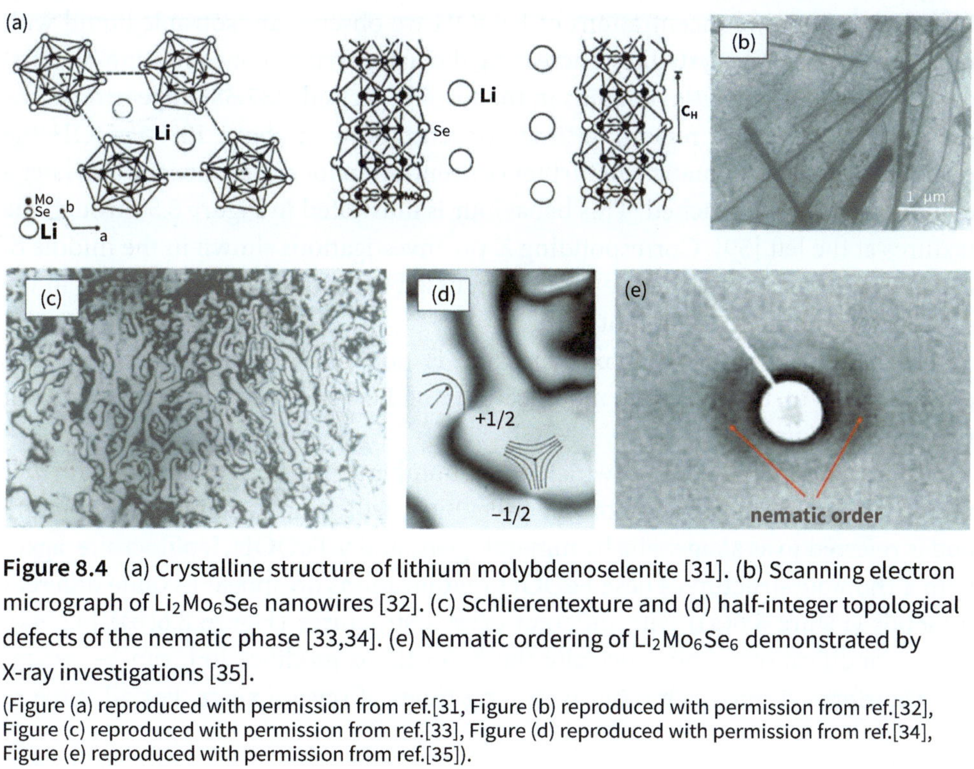

Figure 8.4 (a) Crystalline structure of lithium molybdenoselenite [31]. (b) Scanning electron micrograph of $Li_2Mo_6Se_6$ nanowires [32]. (c) Schlierentexture and (d) half-integer topological defects of the nematic phase [33,34]. (e) Nematic ordering of $Li_2Mo_6Se_6$ demonstrated by X-ray investigations [35].

(Figure (a) reproduced with permission from ref.[31, Figure (b) reproduced with permission from ref.[32], Figure (c) reproduced with permission from ref.[33], Figure (d) reproduced with permission from ref.[34], Figure (e) reproduced with permission from ref.[35]).

The structure of lithium molybdenoselenite is shown in Figure 8.4(a) in projection along two different directions (projection into the ab-plane and the ac-plane), illustrating the quasi 1D-structure of Mo_6Se_6 chains [31], thus the formation of nanowires as they are also seen in SEM, depicted in Figure 8.4(b)[32]. These will then spontaneously orient and form a lyotropic nematic phase with typical Schlierentextures as in Figure 8.4(c) [33,34]. A closer look nicely illustrates the $s=\pm 1/2$ strength topological defects (Figure 8.4(d)). The nematic ordering of $Li_2Mo_6Se_6$ has also been demonstrated via X-ray investigations [35].

8.1.4 Iron oxyhydroxide, FeOOH, (goethite, akageneite, lepidocroite)

Another well investigated lyotropic liquid crystal is the FeOOH sol (iron oxyhydroxide), which again goes back to Zocher and his co-workers [36], who synthesized this via the hydrolysis of iron chloride in an ammonia solution. This is a process which can apparently take several months, but eventually yields a precipitate which can be used directly or by re-dispersion via heating. The building blocks of the liquid crystalline phases are FeOOH crystallites that generally have an approximate size of about 100 nm in width and are several hundred nanometres long, thus a typical nanorod-like system.

The interesting feature of FeOOH lyotropic systems is the fact that here not only nematic, but also smectic phases can occur, as demonstrated in Figure 8.5 for goethite, which is a modification named α-FeOOH. As for other colloidal liquid crystals, at low concentrations of FeOOH we observe an isotropic liquid with randomly dispersed crystallites. Increasing the nanoparticle concentration one first observes ordered nematic domains in the form of tactoids [37,38] appearing at the start of the two-phase region, until at concentrations of about 10% FeOOH the phase is completely nematic. On further concentration increase a smectic and later a columnar phase are reached. This behaviour is illustrated in Figure 8.5 through the textures at the left [39]. Corresponding X-ray investigations shown in the middle of Figure 8.5[40] provide evidence for the discussed behaviour, which is shown in the schematic structures at the right [41].

Yet, the situation for iron oxyhydroxide is not quite as straightforward as it appears, due to the different modifications of FeOOH. This is discussed with reference to Figure 8.6.

Besides goethite, α-FeOOH, which has a rhombic structure, there are two further modifications of iron oxyhydroxide: β-FeOOH, which has a tetragonal structure and is referred to as akageneite by mineralogists; and γ-FeOOH, lepidocroite, again with a rhombic structure. These FeOOH modifications do appear as quite different minerals (Figure 8.6(a)) with different crystal structures (Figure 8.6(b)) [42–44]. These modifications, and especially the β- to the α-modification, can be gradually transformed into each other over long times of several years. For all three of

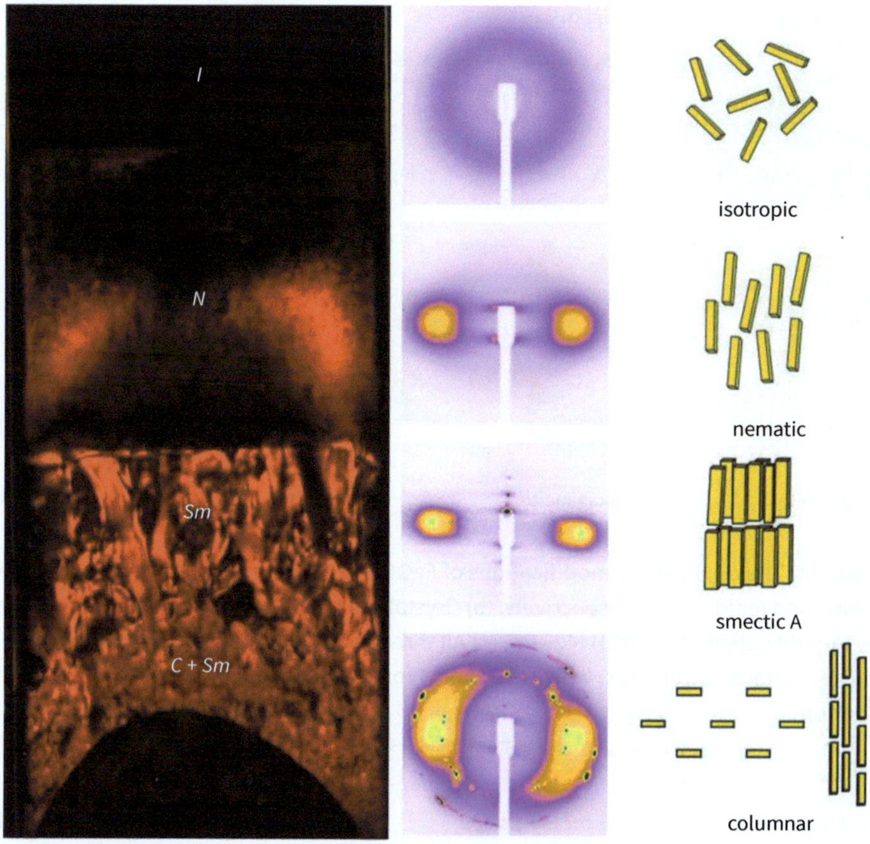

Figure 8.5 Goethite, α-FeOOH, exhibits several different liquid crystal phases. Left: textures of the isotropic, nematic, smectic A, and columnar phases for increasing concentration. Middle: respective small angle X-ray scattering diagrams of the phases observed; and right: the corresponding schematic structure.
(Reproduced with permission from ref.[41]).

the FeOOH-modifications in SEM one observes the typical rod-shaped, elongated, or needle-like crystallites (Figure 8.6(c)) that one would expect for colloidal liquid crystals [44–46].

As mentioned, β-FeOOH liquid crystals had already been discovered about a century ago, possibly also because of their optical properties, which exhibit strikingly brilliant interference colours [13,36,47]. The smectic phase is formed from nanorods, of which an example is depicted by SEM in Figure 8.7(a). The width of these nanorods is about 65 nm with a square shape that was demonstrated by atomic force microscopy (AFM) [48] and the length is approximately 350 nm. These nanorods are thus elongated cuboids and stack orthogonally, with their long axis approximately paralleled, like a smectic A phase, as indicated in the schematic of Figure 8.7 and in part (b) of the figure which shows a colloidal (liquid) crystal in SEM, viewing at an oblique angle. The smectic layer structure is also demonstrated by the typical smectic layer steps seen in microscopy (Figure 8.7(c)) [49]. The cuboid

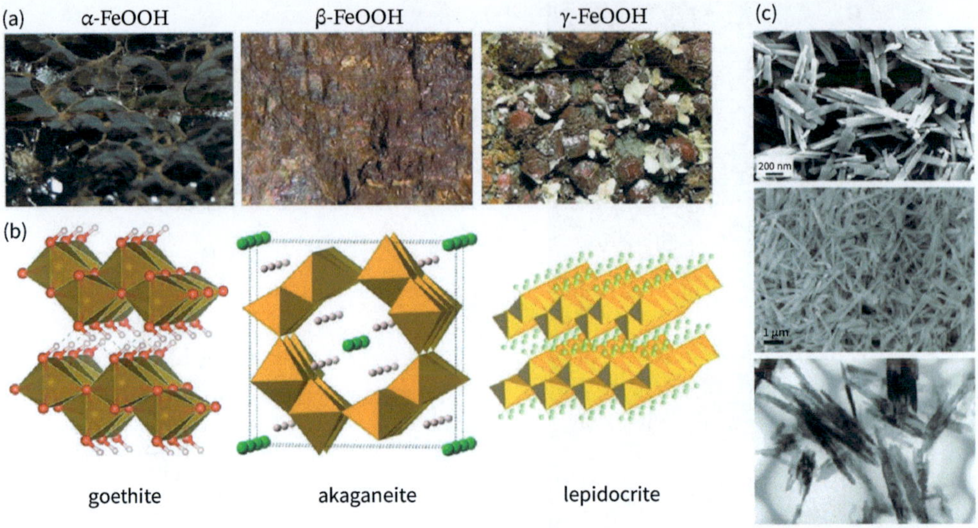

Figure 8.6 (a) The α-, β-, and γ-modifications of FeOOH are different minerals called goethite, akageneite, and lepidocroite, respectively. (b) Crystalline structures of the three different minerals/modifications of FeOOH [42–44]. (c) All three FeOOH modifications exhibit needle-like crystallite structures, which are shown for α- [45], β- [46], and γ-FeOOH [44] from top to bottom.

(Figure (b1,b2,b3), reproduced with permission from ref.[42–44], Figure (c2) reproduced with permission from ref.[44–46]).

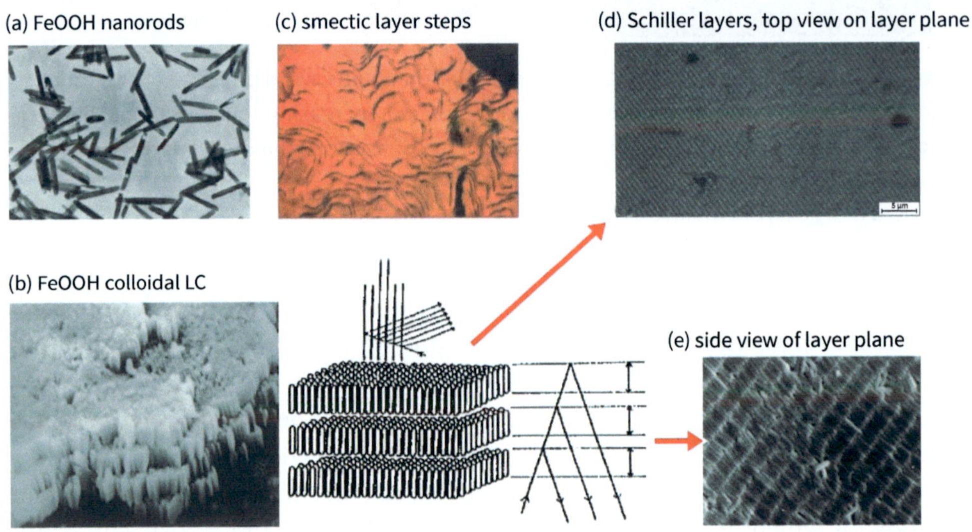

Figure 8.7 (a) SEM of β-FeOOH nanorods [50]. (b) At higher concentrations the smectic A phase is observed, as shown in the atomic force microscopic (AFM) image with nanorods being packed parallel and arranged in layers [48]. (c) Depicts a typical texture of smectic steps observed for free surfaces of smectic A droplets [49]. The schematic figure in the middle shows the two periodicities of Schiller layers [50], which are demonstrated in (d) as a top view of the smectic layer plane [50], and in (e) as a side view [48].

(Figure (a) and (d) reproduced with permission from ref.[50], Figure (b) reproduced with permission from ref.[48], Figure (c) reproduced with permission from ref.[49]).

structure of the building blocks give rise to two periodicities, the one of the equidistant line pattern shown in Figure 8.7(d), and that of the smectic layer thickness, shown in Figure 8.7(e).

The periodicity of the line pattern in the horizontal smectic layer configuration gives rise to an iridescent colour play depending on the angle of incidence of the incoming light. This is what is called interference in the Schiller layers. In a macroscopic sample horizontal layers and vertical layers can coexist, with defects separating both domains. Smectic layer undulations, deformations, and dislocations have also been observed [48].

8.1.5 Imogolite, $(OH)_3Al_2O_3SiOH$

Imogolite is a naturally occurring nanotube with an inner diameter of approximately 10Å while the outer diameter is about 25Å. The mineral is a hydrated aluminium silicate with the formula $(OH)_3Al_2O_3SiOH$ that consists of a gibbsite tube, $Al(OH)_3$, with Al-OH groups on the outside and Si-OH groups on the inside of the hollow cylinder. Imogolite was first discovered in volcanic ash but can now also be made synthetically. The structure of the imogolite nanotubes is shown in Figure 8.8(a) [51], together with an SEM image of synthetically grown imogolite nanotubes in Figure 8.8(b) [52].

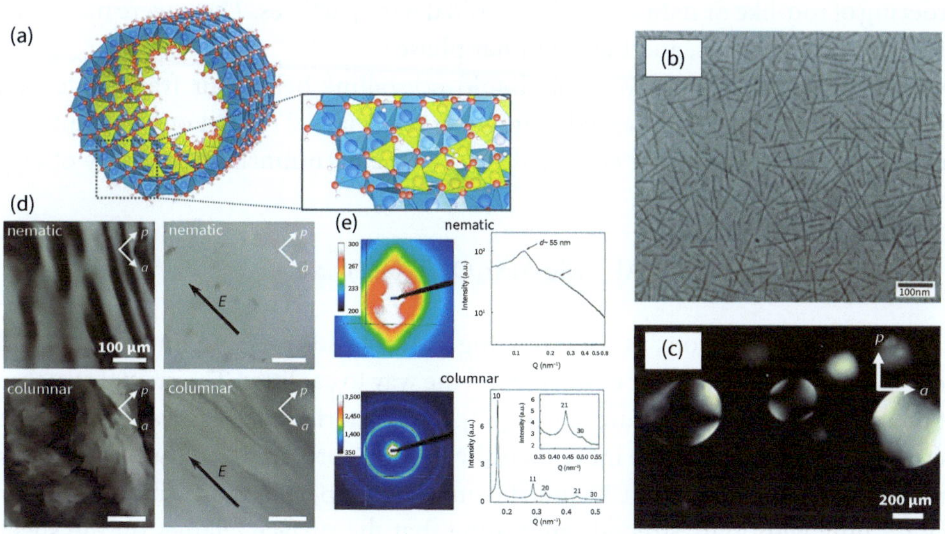

Figure 8.8 (a) Crystal structure of the imogolite nanotubes [51] and (b) a scanning electron micrograph of synthetically grown imogolite tubes [52]. (c) illustrates the growth of the nematic phase via tactoids [55], while in (d) the transition from the nematic to the columnar phase is illustrated without and with application of an electric field E to aid orientation. (e) The nematic and columnar structures can also be evidenced via X-ray diffraction [55].

(Figure (b) reproduced with permission from ref.[52], Figure (c) reproduced with permission from ref.[55], Figure (d) reproduced with permission from ref.[55], Figure (e) reproduced with permission from ref.[55]).

Given their structure, the imogolite nanotubes are easily dispersed in a polar liquid and form a nematic phase [53]. Effects of molecular weight and polydispersity have been studied by Donkai et al. [54] and are well described by Onsager's theory. As for many other rod-like lyotropic systems, the imogolite nematic phase also separates from the isotropic liquid via the growth of tactoids, as shown in Figure 8.8(c)[55]. In fact, at larger, but still small, imogolite concentrations the transition to a columnar phase can be observed (Figure 8.8(d)). Both the nematic and the columnar phases can be oriented by application of an electric field, and the respective structures have been revealed by X-ray diffraction (Figure 8.8(e)) [55]. For a further overview on the structure and properties of imogolite, we may refer the reader to a book chapter by P. Davidson and I. Dozov [56].

8.2 Plate-like mineral materials

8.2.1 Montmorillonites: bentonite and laponite

Just like in thermotropic liquid crystals, the observed liquid crystalline phases can depend on the structure of the molecular building block, for example calamitic phases for rod-like molecules and discotic phases for plate-like or disc-like molecules; a similar behaviour can also be seen for lyotropic phases. In the previous section we introduced a few examples of lyotropic liquid crystals formed by ordering of rod-like or nanotube-like colloidal nanoparticles. These were mainly the nematic, the smectic A, and the columnar phases.

We will now introduce lyotropic liquid crystalline behaviour from plates and discs, mainly the phases observed in liquid crystal clays. The first one to mention is bentonite, a whitish powder (Figure 8.9(a)) of the montmorillonite family of clays with a structural formula of:

$$\mathrm{Na_x\left(Al_{2-x}Mg_x\right)(Si_4O_{10})(OH)_2, zH_2O.}$$

It is found in applications as wide-ranging as food additives and nutritional products, to face masks and other cosmetics, all the way to cat litter. The crystal structure of bentonite is shown in Figure 8.9(b) [57]. The material swells as water penetrates between the layers while the individual sheets are approximately 1 nm in thickness and have a diameter between 30 and 300 nm. The properties of the suspension vary significantly with ionic strength, indicating that the surface charges on the sheets play an important role. An SEM image of bentonite shown in Figure 8.9(c) indicates the layered structure of this clay [58].

Langmuir [59] had already observed birefringence in clay suspensions in 1938 and the separation into an isotropic and a nematic phase in the two-phase region. Textures between crossed polarizers were studied by Emmerson [60] and Gabriel et al. [61] and clearly show nematic behaviour (Figure 8.9(d)). In the transition behaviour between different phases, temperature does not play any significant role

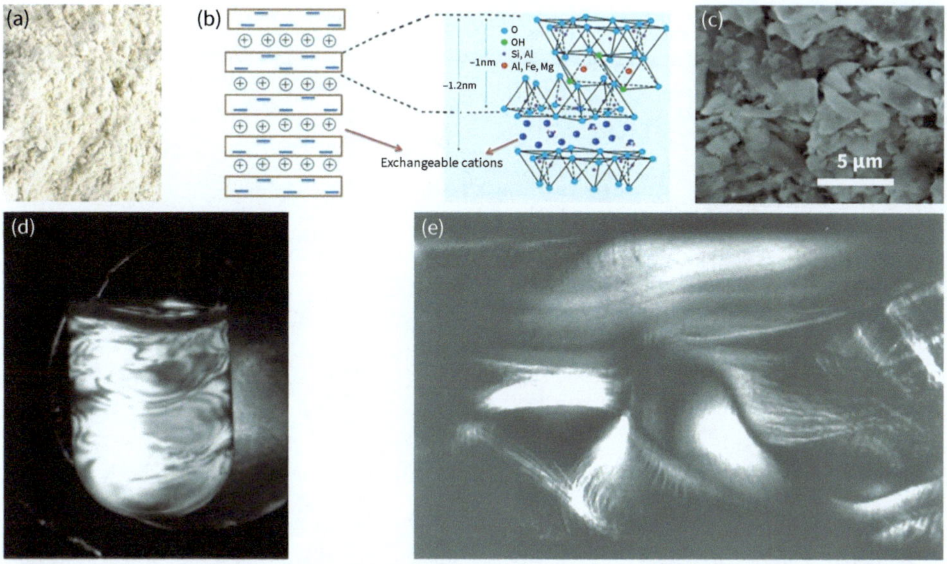

Figure 8.9 (a) Bentonite is a white powder with a structure shown schematically in (b) [57]. (c) shows an SEM image of the plate-like structure of bentonite [58]. (d) and (e) depict nematic textures in bulk and thin layers of bentonite plates in solution [61].

(Figure (a) and (b) reproduced with permission from ref.[57], Figure (c) reproduced with permission from ref.[58], Figure (d) and (e) reproduced with permission from ref.[61]).

and the behaviour is solely governed by clay concentration. At very low concentrations a liquid suspension is observed, which turns into an isotropic gel for increasing concentration. Further increase of clay nanodiscs then results in the formation of a birefringent, nematic gel (Figure 8.9(e)). This behaviour is somewhat different as discussed above for vanadium pentoxide.

Another member of the montmorillonite family is laponite with a chemical structure of:

$$\left[(Na_2Ca)_{x/2}\left(Li_xMg_{3-x}\right)(Si_4O_{10})(OH)_2,\ z\ H_2O\right].$$

Laponite is a synthetic clay which finds applicational use in personal care products where it is used to change the rheological properties. The crystal structure of a laponite sheet is depicted in Figure 8.10(a) [62], with an SEM image of Figure 8.10(b) [63] clearly showing the plate-like structure of the clay. Also here, one can observe the texture of a birefringent nematic lyotropic gel in a test tube preparation (Figure 8.10(c)) as well as a thin film (Figure 8.10(d)) [61]. Small-angle X-ray scattering (SAXS) has shown that the director of the discs (i.e., the disc normal) follows the typical Maier–Saupe behaviour and fitting of the X-ray data led to an orientational order parameter of $S=0.55\pm0.05$ [64], which falls approximately in the range of order parameters that are also observed for thermotropic nematics. Furthermore, it was shown that correlation between centres of mass of the individual discs is extremely weak.

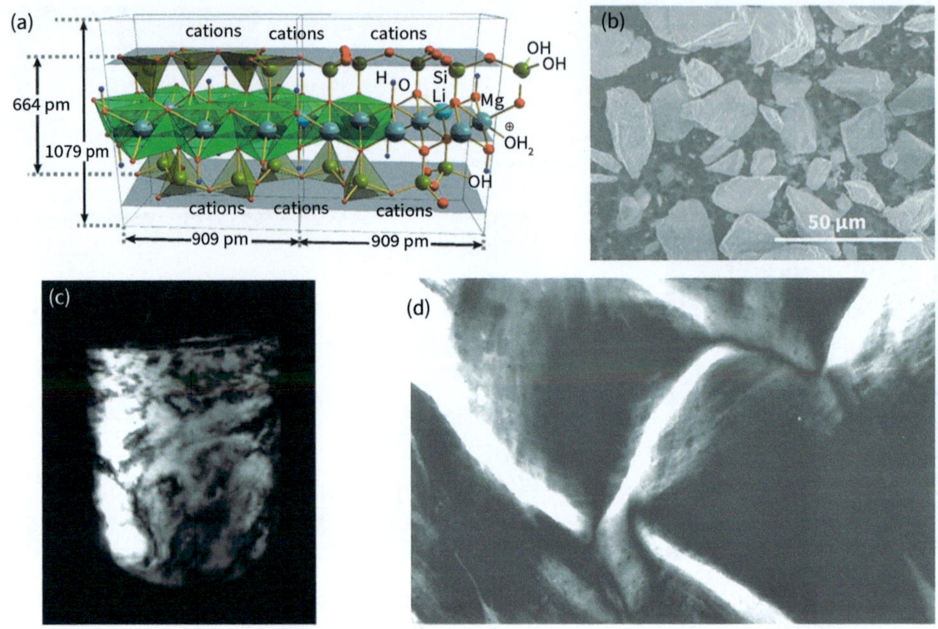

Figure 8.10 (a) Structure of laponite [62] and (b) SEM image demonstrating the plate-like building blocks of the nematic clay phase [63]. Nematic textures in (c) and (d) again show a test tube bulk preparation and a thin film, respectively, of a laponite gel [61].
(Figure (a) reproduced with permission from ref.[62], Figure (c) and (d) reproduced with permission from ref.[61]).

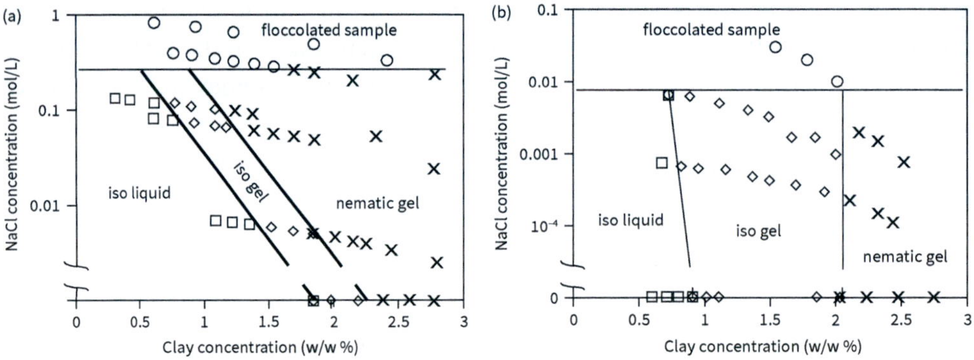

Figure 8.11 Phase diagram of montmorillonite liquid crystals (a) bentonite and (b) laponite, as a function of concentration and ionic strength [34].
(Figure (a) and (b) reproduced with permission from ref.[34]).

We will briefly discuss the phase behaviour of the two montmorillonite clays, bentonite (Figure 8.11(a)) and laponite (Figure 8.11(b), as a function of clay concentration and ionic strength [34]. It can be seen that the phase behaviour is very similar, as one might expect. The transitions are equivalent passing from an isotropic liquid at low clay concentration to an isotropic gel at approximately 1–1.5% by weight and then to a nematic gel at higher concentrations of about 1–2 weight%. In the case of bentonite a dependence on ionic strength is observed, with the transition occurring at lower clay concentrations for higher ionic strength. In the case of laponite the transitions are practically only governed by the clay concentration.

8.2.2 Gibbsite, $Al(OH)_3$

Gibbsite, $Al(OH)_3$, is the classical Onsager lyotropic liquid crystal. It is formed by hexagonal plates of the crystalline structure shown in Figure 8.12(a) [65] and follows Onsager's original theory [66,4], which predicts the isotropic to nematic phase transition for plate-like particles, here with thickness and diameter of 15 nm and 150 nm, respectively (the exemplary SEM image of a gibbsite plate in Figure 8.12(b) taken from ref. [67] is somewhat smaller), as a function of increasing particle concentration (Figure 8.12(c)) [68].

Starting with an isotropic suspension of low concentrations of gibbsite plates in toluene, increasing the volume fraction to approximately 0.16–0.17 leads to phase separation and a two-phase region where the isotropic and the nematic phase coexist. The latter is clearly detectable through its birefringence. On further increasing the gibbsite concentration, the relative fraction between isotropic and nematic phase shifts towards the liquid crystal until at the end of the two-phase region the whole sample is in the nematic state. It should be mentioned that the first time this behaviour was clearly observed was about five decades after Onsager's theory, when Van der Kooij and Lekkerkerker [69] sterically stabilized the gibbsite plates by grafting them with a layer of poly-isobutene. This experiment is indeed of significant importance, because the grafting of the polymer onto the gibbsite particles demonstrates that the interaction is purely hard core and not dispersion interactions (van der Waals) or electrostatic. The fact that no gelation occurs further shows that the formation of the nematic phase is indeed thermodynamic in nature. The observed

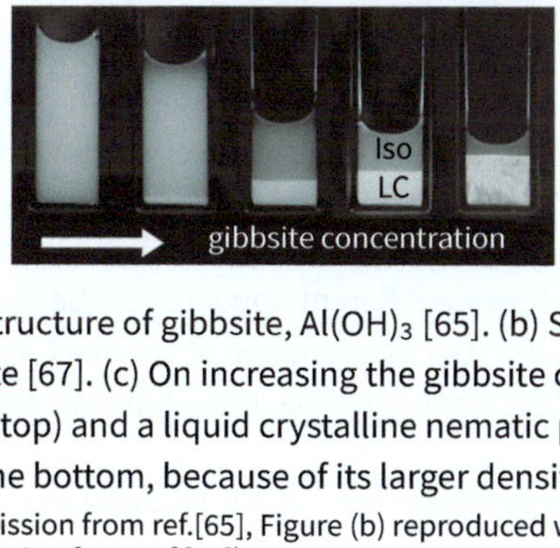

Figure 8.12 (a) Crystalline structure of gibbsite, $Al(OH)_3$ [65]. (b) Scanning electron micrograph of a gibbsite plate [67]. (c) On increasing the gibbsite concentration, phase separation into an isotropic (top) and a liquid crystalline nematic phase (bottom) is observed. The LC phase is situated at the bottom, because of its larger density.

(Figure (a) reproduced with permission from ref.[65], Figure (b) reproduced with permission from ref.[67], Figure (c) reproduced with permission from ref.[68]).

width of the two-phase region fits well with theory and was also confirmed by the computer simulations of Bates and Frenkel [70].

8.2.3 $H_3Sb_3P_2O_{14}$

A quite unconventional lyotropic material exhibiting a lamellar phase, is $H_3Sb_3P_2O_{14}$. This was obtained by exfoliation of the low-dimensional, quasi-2D $H_3Sb_3P_2O_{14}$ material in water [71]. The lamellar phase consists of large covalently bonded sheets and was identified via SAXS on shear-oriented samples [72]. At its largest extension the lamellar period was found to be 225 nm, which corresponds to the largest swelling of the structure. The complete phase sequence of $H_3Sb_3P_2O_{14}$ as a function of increasing concentration is given by isotropic–nematic–lamellar, but it is also strongly dependent on ionic strength [73].

Polarizing microscopy at a $H_3Sb_3P_2O_{14}$ concentration of 0.57% by weight exhibited the coexistence of three phases, as demonstrated in Figure 8.13(a) [72] with the top being a homogeneous suspension of nanoparticles in the isotropic phase.

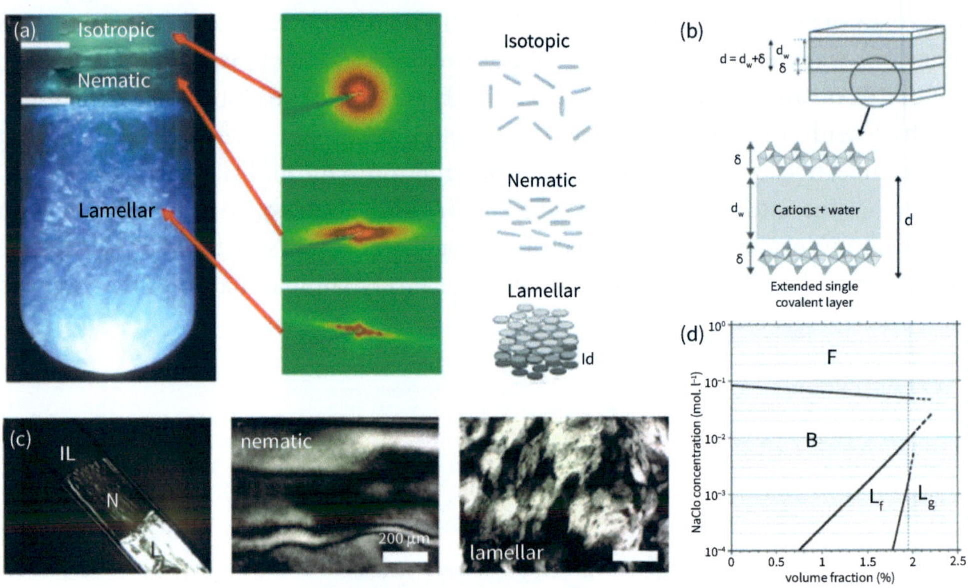

Figure 8.13 (a) Formation of a lamellar phase in a $H_3Sb_3P_2O_{14}$ with a test tube preparation exhibiting the separation into the isotropic liquid (top), the liquid crystalline nematic (middle), and lamellar (bottom) phases, together with the corresponding X-ray diffraction diagrams and the schematic structures [73]. (b) Model structure of the lamellar phase [71]. (c) Textures of a thin-film preparation clearly showing different appearances between nematic and lamellar [73]. (d) Phase diagram of $H_3Sb_3P_2O_{14}$ as a function of volume fraction concentration and ionic strength [71], where F is the flocculation regime, B the biphasic regime, L_f the lamella fluid, and L_g the lamellar gel phase.

(Figure (a) and (c) reproduced with permission from ref.[73], Figure (b) and (d) reproduced with permission from ref.[71]).

The middle phase is nematic and reacts to the application of small magnetic fields, while the bottom phase displays the lamellar phase, which does not respond to magnetic fields due to its high viscosity. SAXS shows the typical corresponding scattering profiles with the lamellar phase giving rise to several reflections. The proposed structure of the lamellar phase is depicted in Figure 8.13(b) [71]. The transition between nematic and lamellar shows a clear change of texture which is demonstrated in Figure 8.13(c) [73]. The phase diagram varying concentration and ionic strength is given in Figure 8.13(d) [71], where the biphasic region, B, transforms into a fluid lamellar phase and at higher concentrations into the lamellar gel phase. Flocculation, F, is observed for high ionic strength, i.e., large concentrations of NaCl. (The nematic phase is not shown here.)

8.3 Mixtures of mineral rods and discs

The miscibility and phase behaviour of thermotropic mixtures of rods and discs has been a long-standing question, especially as this is closely linked to the possible occurrence of biaxial nematic phases. Simply mixing a calamitic with a discotic mesogen does not lead to success, because phase separation is observed very quickly. This has also been suggested by theory [74] through entropically driven de-mixing, as well as computer simulations [75]. It was later proposed to prohibit phase separation [76], which can be accomplished by covalently connecting rod-shaped with disc-shaped molecules to form a rod–disc dimer [77,78]. In some cases full miscibility seems to have been observed [79].

Similar experiments can also be carried out for mineral liquid crystals, simply by mixing rod-shaped colloidal particles with disc-shaped ones, as for example in the case of boehmite rods with gibbsite plates, which is structurally shown in Figure 8.14(a) [65]. The respective experiment was performed by van der Kooij and Lekkerkerker [80]. Depending on the volume fraction of rods and discs, they found phase-separated mixtures, one of which showed coexistence between the isotropic phase (I), a rod-rich phase (N+) and a disc-rich phase (N−), from which the transmission electron micrographs (TEM) were achieved that are shown in Figure 8.14(b). Varying the volume fractions, three typical texture scenarios can be observed, as depicted in Figure 8.14(c). In that figure, (i) shows an isotropic phase in coexistence with a rod-like nematic phase, i.e., (I+N+). At the other end of the phase diagram a similar situation is seen in (iii) with an isotropic phase in coexistence with a disc-like nematic phase, i.e., (I+N−). In the middle of the phase diagram (ii) coexistence between three phases, the isotropic (top), a nematic rod-like (middle), and a nematic plate-like phase (bottom) is observed. This is also summarized in Figure 8.14(d). At higher concentrations, above $\Phi \approx 0.3$, the phase diagram becomes even richer, with columnar and smectic phases present [80], which do not need to be discussed here. In summary, the authors find that a biaxial phase becomes unstable and de-mixing occurs into two uniaxial nematic and the isotropic phase.

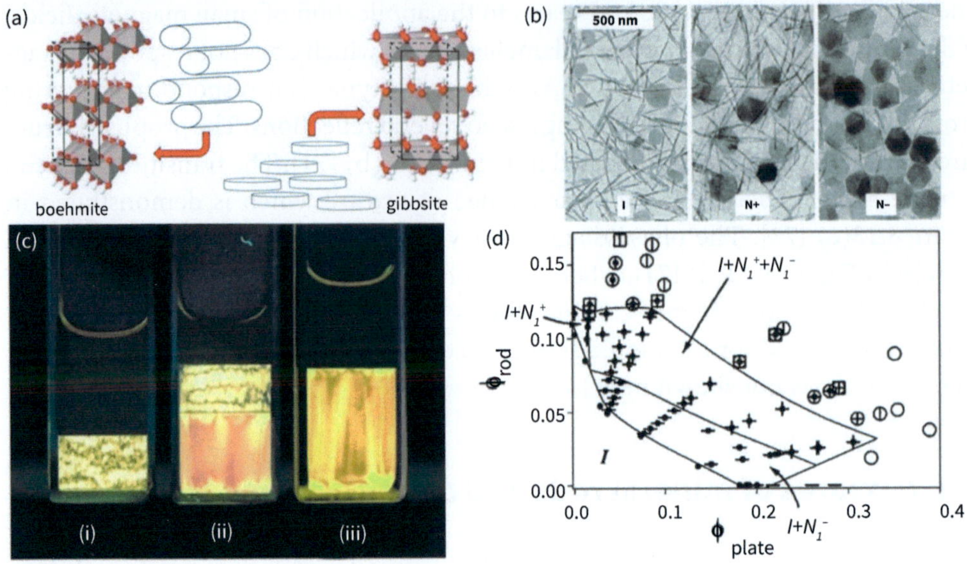

Figure 8.14 (a) Mixtures of rods and discs can be achieved through mixing boehmite with gibbsite particles [65]. (b) Mixtures of boehmite rods with gibbsite plates shown for increasing gibbsite concentration via scanning electron microscopy [80]. (c) (i) coexistence between the isotropic and the rod-rich phase, (ii) between the isotropic, the rod-rich, and the plate-rich phase, and (iii) between the isotropic and the plate-rich phase, on increasing the plate concentration [80]. (d) Phase diagram of the rod–plate system [80], as discussed in the text. (Figure (a) reproduced with permission from ref.[65], Figure (b), (c) and (d) reproduced with permission from ref.[80]).

For more detailed information on mineral and inorganic liquid crystals one may refer to a number of review articles [34,68,81–85].

References

[1] Q. Majorana (1902). Sur la birefringence magnetique. Comptes Rendus Hebd. *Seances Acad. Sci.*, 135, 159.

[2] A. Cotton and H. Mouton (1905). Sur le phenomene de Majorana. *Comptes Rendus Acad. Sci.*, 141, 317.

[3a] W. Maier and A. Saupe (1958). Eine einfache molekulare Theorie des nematischen kristallinflüssigen Zustandes. *Zeitschrift für Naturforsch. A*, 13, 564–566.

[3b] W. Maier and A. Saupe (1959). Eine einfache molekular-statistische Theorie der nematischen kristallinflüssigen Phase. Teil II. *Zeitschrift für Naturforsch. A*, 14, 882.

[3c] W. Maier and A. Saupe (1960). Eine einfache molekular-statistische Theorie der nematischen kristallinflüssigen Phase. Teil III. *Zeitschrift für Naturforsch. A*, 15, 287.

[4] L. Onsager (1949). The effects of shape on the interaction of colloidal particles. *Ann. N. Y. Acad. Sci.*, 51, 627–659.

[5] A. Stroobants, H. N. W. Lekkerkerker, and T. Odijk (1986). Effect of electrostatic interaction on the liquid crystal phase transition in solutions of rodlike polyelectrolytes. *Macromolecules*, 19, 2232–2238.

[6] G. J. Vroege and H. N. W. Lekkerkerker (1992). Phase transitions in lyotropic colloidal and polymer liquid crystals. *Rep. Prog. Phys.*, 55, 1241–1309.

[7] H. Freundlich (1916). Die Doppelbrechung des Vanadinpentoxydsols. *Ber. Bunsenges. Phys. Chem.*, 22, 27–33.

[8] H. Zocher (1921). Über Sole mit nichtkugeligen Teilchen. *Z. Phys. Chem.*, 98U, 293–337.

[9] E. Londero and E. Schröder (2011). Vanadium pentoxide (V_2O_5): A van der Waals density functional study. *Computer Physics Communications*, 182, 1805–1809.

[10] M. Dhananjaya, N. Guru Prakash, A. Lakshmi Narayana, and O. M. Hussain (2018). Microstructural and supercapacitive properties of one-dimensional vanadium pentoxide nanowires synthesized by hydrothermal method. *Applied Physics A*, 124, 185.

[11] O. Pelletier, P. Davidson, C. Bourgaux, and J. Livage (1999). Liquid crystalline aqueous suspensions of V_2O_5 ribbons: From repulsive sols to attractive gels. *Progr. Colloid Polym. Sci.*, 112, 121–125.

[12] H. Freundlich, F. Stapelfeldt, and H. Zocher (1924). Quantitative Untersuchungen am Vanadinpentoxysol. *Z. Phys. Chem.*, 114U, 161–189.

[13] H. Zocher (1925). Über freiwillige Strukturbildung in Solen. *Z. Anorg. Allg. Chem.*, 147, 91–110.

[14] A. V. Kaznacheev, M. M. Bogdanov, and S. A. Taraskin (2002). The nature of prolate shape of tactoids in lyotropic inorganic liquid crystals. *J. Exp. Theo. Phys.*, 95, 57–63.

[15] A. V. Kaznacheev, M. M. Bogdanov, and A. S. Sonin (2003). The influence of anchoring energy on the prolate shape of tactoids in lyotropic inorganic liquid crystals. *J. Exp. Theo. Phys.*, 97, 1159–1167.

[16] P. Prinsen and P. van der Schoot (2003). Shape and director-field transformation of tactoids. *Phys. Rev. E*, 68, 021701.

[17] L. Metselaar, I. Dozov, K. Antonova, E. Belamie, P. Davidson, J. M. Yeomans, and A. Doostmohammadi (2017). Electric-field-induced shape transition of nematic tactoids. *Phys. Rev. E*, 96, 022706.

[18] A. A. Verhoeff, R. H. J. Otten, P. van der Schoot, and H. N. W. Lekkerkerker (2011). Magnetic field effects on tactoids of plate-like colloids. *J. Chem. Phys.*, 134, 044904.

[19] X. Commeinhes, P. Davidson, C. Bourgaux, and J. Livage (1997). Orientation of liquid-crystalline suspensions of vanadium pentoxide ribbons by a magnetic field. *Adv. Mater.*, 9, 900–903.

[20] O. Pelletier, P. Davidson, C. Bourgaux, C. Coulon, S. Regnault, and J. Livage (2000). A detailed study of the synthesis of aqueous vanadium pentoxide nematic gels. *Langmuir*, 16, 5295–5303.

[21] J. Livage, O. Pelletier, and P. Davidson (2000). Vanadium pentoxide sol and gel mesophases. *J. Sol-Gel Sci. Technol.*, 19, 275–278.

[22] P. Davidson, A. Garreau, and J. Livage (1994). Nematic colloidal suspensions of V_2O_5 in water—or Zocher phases revisited. *Liq. Cryst.*, 16, 905–910.

[23] O. Pelletier, P. Sotta, and P. Davidson (1999). Deuterium nuclear magnetic resonance study of the nematic phase of vanadium pentoxide aqueous suspensions. *J. Phys. Chem. B*, 103, 5427–5433.

[24] O. Pelletier, P. Davidson, C. Bourgaux, and J. Livage (1999). The effect of attractive interactions on the nematic order of V_2O_5 gels. *Europhys. Lett.*, 48, 53–59.

[25] O. Pelletier, C. Bourgaux, O. Diat, P. Davidson, and J. Livage (1999). A biaxial nematic gel phase in aqueous vanadium pentoxide suspensions. *Eur. Phys. J. B*, 12, 541–546.

[26] J. Livage (1991). Vanadium pentoxide gels. *Chem. Mater.*, 3, 578–593.

[27] H. Zocher and C. Török (1960). Neuere Beiträge zur Kenntnis der Taktosole. *Kolloid Zeitschrift*, 170, 140–144.

[28] L. Bohaty, O. Lux, H.-J. Eichler, H. Rhee, A. A. Kaminskii, and P. Becker (2021). Stimulated Raman scattering (SRS) in α-AlOOH (diaspore). *Cryst. Res. Technol.*, 56, 2100055.

[29] A. Hermawan, H. W. Son, Y. Asakura, T. Mori, and S. Yin (2018). Synthesis of morphology controllable aluminum nitride by direct nitridation of γ-AlOOH in the presence of N_2H_4 and their sintering behavior. *J. Asian Ceramic Soc.*, 6, 63–69.

[30] M. Potel, R. Chevrel, and O. Fischer (1979). New clusters in the ternary chalcogenides derived from molybdenum (II)—Mo_9, Mo_{12}, and $(Mo_3)^1\infty$. *Compt. Rend. Hebd. Seances Acad. Sci. C*, 288, 429–432.

[31] M. Potel, R. Chevrel, M. Sergent, J. C. Armici, M. Decroux, and O. Fischer (1980). New pseudo-one-dimensional metals: $M_2Mo_6Se_6$ (M = Na, In, K, TI), $M_2Mo_6S_6$ (M = K, Rb, Cs), $M_2Mo_6Te_6$ (M = In, TI). *J. Solid State Chem.*, 35, 286–290.

[32] J. M. Tarascon, F. J. DiSalvo, C. H. Chen, P. J. Carroll, M. Walsh, and L. Rupp. (1985). First example of monodispersed $(Mo_3Se_3)^1\infty$ clusters. *J. Solid State Chem.*, 58, 290–300.

[33] P. Davidson, J.-C. Gabriel, A. M. Levelut, and P. Batail (1993). A new nematic suspension based on all-inorganic polymer rods. *Europhys. Lett.*, 21, 317–322.

[34] J.-C. Gabriel and P. Davidson (2000). New trends in colloidal liquid crystals based on mineral moieties. *Adv. Mater.*, 12, 9–20.

[35] P. Davidson, J.-C. Gabriel, A. M. Levelut, and P. Batail (1993). Nematic liquid crystalline mineral polymers. *Adv. Mater.*, 5, 665–668.

[36] H. Zocher and W. Heller (1929). Schillerschichten als Reaktionsprodukte der langsamen Eisenchlorid-Hydrolyse. *Z. Anorg. Allg. Chem.*, 186, 75–96.

[37] K. Coper and H. Freundlich (1937). The formation of tactoids in iron oxide sols. *Trans. Farad. Soc.*, 33, 348–350.

[38] A. L. Mackay (1962). β-ferric oxyhydroxide—akageneite. *Mineral. Mag.*, 33, 270–280.

[39] B. J. Lemaire, P. Davidson, J. Ferre, J.-P. Jamet, D. Petermann, P. Panine, I. Dozov, D. Stoenescuc, and J.-P. Jolivet (2005). The complex phase behaviour of suspensions of goethite (a-FeOOH) nanorods in a magnetic field. *Faraday Discuss.*, 128, 271–283.

[40] E. van den Pol, A. V. Petukhov, D. M. E. Thies-Weesie, D. V. Byelov, and G. J. Vroege (2010). Liquid crystal phase behavior of sterically stabilized goethite. *J. Colloid Interface Sci.*, 352, 354–358.

[41] H. N. W. Lekkerkerker and G. J. Vroege (2013). Liquid crystal phase transitions in suspensions of mineral colloids: new life from old roots. *Phil. Trans. Royal Soc. A*, 371, 20120263.

[42] A. A. Shubin, V. Y. Kovalskii, S. Ph. Ruzankin, I. L. Zilberberg, V. N. Parmon, F. N. Tomilin, and P. V. Avramov (2021). The Fe^{IV} O. oxyl unit as a key intermediate in water oxidation on the Fe^{III} hydroxide: DFT predictions. *Int. J. Quantum Chem.*, 121, e26610. https://doi.org/10.1002/qua.26610

[43] K. M. Peterson, P. J. Heaney, and J. E. Post (2018). Evolution in the structure of akaganeite and hematite during hydrothermal growth: an in situ synchrotron X-ray diffraction analysis. *Powder Diffraction*, 33, 287–297.

[44] J.-P. Jolivet, C. Chaneac, and E. Tronc (2004). Iron oxide chemistry: From molecular clusters to extended solid networks. *Chem. Commun.*, 482, 481–487.

[45] F. Agresti, V. Zin, S. Barison, E. Sani, M. Meucci, L. Mercatelli, L. Nodari, S. Rossi, S. Bobboc, and M. Fabrizioa (2017). NIR transmittance tuneability under a magnetic field of colloidal suspensions of goethite (α-FeOOH) nanorods. *RSC Adv.*, 7, 12429.

[46] M. Yang, Y.-X. Li, M. Jiang, P.-H. Li, S.-H. Chen, J.-H. Liu, C.-H. Lin, X.-J. Huang, and W.-Q. Liu (2020). Identifying phase-dependent electrochemical stripping performance of FeOOH nanorod: Evidence from kinetic simulation and analyte–material interactions. *Small*, 16, 1906830.

[47] W. Heller (1935). Distances between the colloidal particles in the bright layers of certain iron oxide soles. Comptes Rendus Hebd. *Seanc. Acad. Sci.*, 201, 831–833.

[48] H. Maeda and Y. Maeda (1996). Atomic force microscopy studies for investigating the smectic structures of colloidal crystals of β-FeOOH. *Langmuir*, 12, 1446–1452.

[49] H. Zocher and C. Török (1967). Crystals of higher order and their relation to other superphases. *Acta Cryst.*, 22, 751–755.

[50] Y. Maeda and S. Hachisu (1983). Schiller layers in β-ferric oxyhydroxide sol as an order-disorder phase separating system. *Colloids and Surfaces*, 6, 1–16.

[51] E. Paineau (2018). Imogolite nanotubes: A flexible nanoplatform with multipurpose applications. *Applied Sciences*, 8, 1921.

[52] H. Yang, C. Wang, and Z. Su (2008). Growth mechanism of synthetic imogolite nanotubes. *Chem. Mater.*, 20, 4484–4488.

[53] K. Kajiwara, N. Donkai, Y. Fujiyoshi, and H. Inagaki (1986). Lyotropic mesophase of imogolite, 2: Microscopic observation of imogolite mesophase. *Macromol. Chem.*, 187, 2895–2907.

[54] N. Donkai, K. Kajiwara, M. Schmidt, and T. Miyamoto (1993). Lyotropic mesophase of imogolite: Molecular weight fractionation and polydispersity effect. *Makromol. Chem. Rapid Commun.*, 14, 611–617.

[55] E. Paineau, M.-E. M. Krapf, M.-S. Amara, N. V. Matskova, I. Dozov, S. Rouziere, A. Thill, P. Launois, and P. Davidson (2016). A liquid-crystalline hexagonal columnar phase in highly-dilute suspensions of imogolite nanotubes. *Nature Communications*, 7, 10271.

[56] P. Davidson and I. Dozov (2016). Liquid crystalline phases of imogolite and halloysite dispersions. In: P. Yuan, A. Thill, and F. Bergaya (eds), *Developments in Clay Science*, Vol.7, Amsterdam: Elsevier, Chapter 13, pp 308–330.

[57] R. S. Hebbar, A. M. Isloor, and A. F. Ismail (2014). Preparation and evaluation of heavy metal rejection properties of polyetherimide/porous activated bentonite clay nanocomposite membrane. *RSC Adv.*, 4, 47240–47248.

[58] J. Ahmad, K. J. Kontoleon, M. Z. Al-Mulali, S. Shaik, M. H. El Ouni, and M. A. El-Shorbagy (2022). Partial substitution of binding material by bentonite clay (bc) in concrete: A review. *Buildings*, 12, 634.

[59] I. Langmuir (1938). The role of attractive and repulsive forces in the formation of tactoids, thixotropic gels, protein crystals and coacervates. *J. Chem. Phys.*, 6, 873–896.

[60] W. W. Emerson (1956). Liquid crystals of montmorillonite. *Nature*, 4544, 1248–1249.

[61] J.-C. P. Gabriel, C. Sanchez, and P. Davidson (1996). Observation of nematic liquid-crystal textures in aqueous gels of smectite clays. *J. Phys. Chem.*, 100, 11139–11143.

[62] M. M. Lezhnina, T. Grewe, H. Stoehr, and U. Kynast (2012). Laponite blue: Dissolving the insoluble. *Angew. Chem. Int. Ed.*, 51, 10652–10655.

[63] Z. Liu, Z. Zhao, X. Jin, L.-M. Wang, and Y. D. Liu (2021). Preparation of cellulose/laponite composite particles and their enhanced electrorheological responses. *Molecules*, 26, 1482.

[64] B. J. Lemaire, P. Panine, J. C. P. Gabriel, and P. Davidson (2002). The measurement by SAXS of the nematic order parameter of laponite gels. *Europhys. Lett.*, 59, 55–61.

[65] R. Demichelis, B. Civalleri, P. D'Arco, and R. Dovesi (2010). Performance of 12 DFT functionals in the study of crystal systems: Al_2SiO_5 orthosilicates and Al hydroxides as a case study. *Int. J. Quantum Chem.*, 110, 2260–2273.

[66] L. Onsager (1942). Anisotropic solutions of colloids. *Phys. Rev.*, 62, 558. See also ref. [4].

[67] F. J. Freij, G. M. Parkinson, and M. M. Reyhani (2004). Atomic force microscopy study of the growth mechanism of gibbsite crystals. *Phys. Chem. Chem. Phys.*, 6, 1049–1055.

[68] Image courtesy of David van der Beek, reproduced from: P. Davidson and J.-P. C. Gabriel (2005). Self-assemblies of anisotropic nanoparticles: Mineral liquid crystals. In: M.-P. Pileni (ed.), *Nanocrystals forming mesoscopic structure*, Weinheim: Wiley-VCH, Chapter 7, pp 173–212. See also: D. van der Beek and H. N. W. Lekkerkerker (2003). Nematic ordering vs. gelation in suspensions of charged platelets. *Europhys. Lett.*, 61, 702–707.

[69] F. M. van der Kooij and H. N. W. Lekkerkerker (1998). Formation of nematic liquid crystals in suspensions of hard colloidal platelets. *J. Phys. Chem. B*, 102, 7829–7832.

[70] M. A. Bates and D. Frenkel (1999). Nematic–isotropic transition in polydisperse systems of infinitely thin hard platelets. *J. Chem. Phys.*, 110, 6553–6559.

[71] J.-C. P. Gabriel, F. Camerel, B. J. Lemaire, H. Desvaux, P. Davidson, and P. Batail (2001). Swollen liquid-crystalline lamellar phase based on extended solid-like sheets. *Nature*, 413, 504–508.

[72] F. Camerel, J. C. P. Gabriel, P. Batail, P. Panine, and P. Davidson (2003). Combined SAXS-rheological studies of liquid-crystalline colloidal dispersions of mineral particles. *Langmuir*, 19, 10028–10035.

[73] P. Davidson, C. Penisson, D. Constantina, and J.-C. P. Gabriel (2018). Isotropic, nematic, and lamellar phases in colloidal suspensions of nanosheets. *PNAS*, 115, 6662–6667.

[74] R. van Roij and B. Mulder (1994). Demixing in a hard rod-plate mixture. *Journal de Physique II*, 4, 1763–1769.

[75] A. Galindo, A. J. Haslam, S. Varga, G. Jackson, A. G. Vanakaras, D. J. Photinos, and D. A. Dunmur (2003). The phase behavior of a binary mixture of rodlike and disclike mesogens: Monte Carlo simulation, theory, and experiment. *J. Chem. Phys.*, 119, 5216–5225.

[76] R. Hashim, G. R. Luckhurst, F. Prata, and S. Romano (1993). Computer simulation studies of anisotropic systems. XXII. An equimolar mixture of rods and discs: A biaxial nematic? *Liq. Cryst.*, 15, 283–309.

[77] J. J. Hunt, R. W. Date, B. A. Timimi, G. R. Luckhurst, and D. W. Bruce (2001). Toward the biaxial nematic phase of low molar mass thermotropic mesogens: Substantial molecular biaxiality in covalently linked rod-disk dimers. *J. Am. Chem. Soc.*, 123, 10115–10116.

[78] K.-U. Jeong, A. J. Jing, B. Monsdorf, M. J. Graham, F. W. Harris, and S. Z. D. Cheng (2007). Self-assembly of chemically linked rod-disc mesogenic liquid crystals. *J. Phys. Chem. B*, 111, 767–777.

[79] P. H. J. Kouwer and G. H. Mehl (2003). Full miscibility of disk- and rod-shaped mesogens in the nematic phase. *J. Am. Chem. Soc.*, 125, 11172–11173.

[80] F. M. van der Kooij and H. N. W. Lekkerkerker (2000). Liquid-crystalline phase behavior of a colloidal rod-plate mixture. *Phys. Rev. Lett.*, 84, 781–784.

[81] A. S. Sonin (1998). Inorganic lyotropic liquid crystals. *J. Mater. Chem.*, 8, 2557–2574.

[82] J.-C. P. Gabriel and P. Davidson (2003). Mineral liquid crystals from self-assembly of anisotropic nanosystems. *Top. Curr. Chem.*, 226, 119–172.

[83] P. Davidson and J.-C. P. Gabriel (2005). Mineral liquid crystals. *Current Opinion in Colloid & Interface Science*, 9, 377–383.

[84] A. S. Sonin, N. A. Churochkina, A. V. Kaznacheev, and A. V. Golovanov (2017). Mineral liquid crystals. *Colloid Journal*, 79, 421–450.

[85] I. Dierking and S. Al-Zangana (2017). Lyotropic liquid crystal phases from anisotropic nanomaterials. *Nanomaterials*, 7, 305.

9
Biological and living lyotropics

9.1 Introduction

The interface between lyotropic liquid crystals and biology is broad [1]. Phospholipids, one of the structural components of the cellular membrane, are amphiphilic molecules. A polar head cohabits with nonpolar parts in the same molecule. This property allows the making of complex structures that separates a polar region from a nonpolar one. The formation of micelles in mixtures of amphiphilic molecules and water above the critical micellar concentration (CMC) is an example of these super-molecular structures. This type of super-molecular structure is observed in the human lipoproteins. Low Density (LDL) and High Density (HDL) Lipoproteins, the main carriers of cholesterol in the circulatory system, are almost spherical and are constituted by a phospholipid shell that encapsulates esters of cholesterol and triglycerides [2,3]. Otherwise, cholesterol, that is not soluble in polar fluids, would not be transported in the blood to the different parts of our body. The typical dimensions of the LDL and HDL are 20 nm and 9 nm, respectively. In this case, the formation of the LDL and HDL does not depend on a particular CMC, and the presence of other proteins attached to them assures the particle stability.

In this chapter we discuss the presence of lyotropic mesophases in systems where the basic units are viruses (tobacco mosaic virus and filamentous phage virus fd), DNA and RNA, actin filaments, and cellulose nanocrystals and derivatives.

9.2 Tobacco mosaic virus (TMV)

Viruses have a striking characteristic that makes them excellent materials to stabilize liquid crystalline mesophases in solutions: their shape and shape anisotropy are the same for all the individuals. TMV was the first virus identified in the history of science, in the late nineteenth century [4]. It is a polyelectrolyte of cylindrical geometry and, at pH 7, is negatively charged (charge density of about $15\, e^-/nm$, where $e^- \sim 1.6 \times 10^{-19}$ *coulombs* is the elementary charge). In the virus structure, proteins are arranged in a helicoidal arrangement, forming a hollow cylinder of length of 300 nm and diameter of 18 nm (Figures 9.1(a) and (b)) [5]. Along the cylindrical protein structure there is a single RNA ribbon, twisted throughout the structure. The hollow tube in the centre of the cylinder has a diameter of 4 nm [6]. In aqueous solutions, cylinders interact, mainly, by electrostatic repulsion forces. The length of the cylinder can be modified by means of chemical (e.g., end-to-end

Lyotropic Liquid Crystals. Ingo Dierking and Antônio Martins Figueiredo Neto, Oxford University Press.
 DOI: 10.1093/9780191924927.003.0009

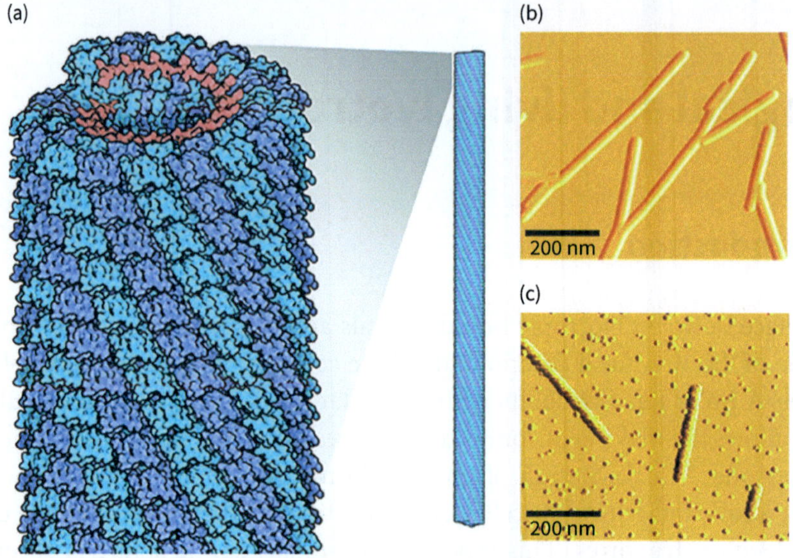

Figure 9.1 (a) Representation of the tobacco mosaic virus, with the RNA genome in red. Image from the RCSB PDB January 2009 Molecule of the Month feature by David Goodsell (http://doi.org/10.2210/rcsb_pdb/mom_2009_1). Tobacco mosaic virus observed by AFM; (b) TMV template immobilized on a Si/SiO_2 substrate; (c) TMV template after 1 cycle of ZnO mineralization.
(Figure (a) RCSB PDB , Figure (b) and (c) reproduced with permission from ref.[5]).

aggregation, to increase the length) and mechanical (e.g., sonication to reduce the length) procedures.

In systems constituted by polymers, or for the viruses discussed in this chapter (typical linear dimension L), the concept of flexibility and rigidity is defined by considering the *persistence length* (D). D gives a length scale of the correlations between independent units. If $D >> L$ the polymer or virus is considered rigid. On the other hand, if $D \sim L$, it is considered flexible.

From the theoretical point of view, liquid crystalline mesophases are expected to be stabilized in aqueous solutions of TMV, as a function of the temperature, the concentration of the viruses, and the ionic strength of the solution [7].

A nematic phase was identified in aqueous (0.05M borate buffer, pH 8.5) solutions of TMV (length of 300 nm), with a typical virus concentration of $0.15 g/cm^3$ [8]. Figure 9.2 shows the partial phase diagram of this solution, with the nematic, isotropic, and coexistence regions. The experimental results are well described by a hard-rod particles model, based on Onsager's theory [9].

Fraden and Maret investigated solutions of monodisperse and polydisperse TMV as a function of the virus concentration, ionic strength, temperature, and applied external magnetic field, in the nematic and isotropic mesophases [10,11]. Their results were shown to be well described by a model based on the Onsager theory, where the viruses interact via repulsive electrostatic force. Attractive forces between viruses were considered to be negligible. The presence of an applied magnetic field ($\vec{H}$) in the isotropic phase of a solution leads to the appearance of a

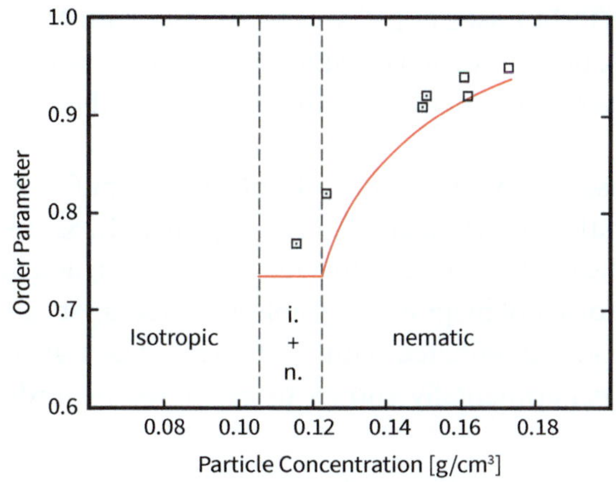

Figure 9.2 Partial phase diagram with the orientational order parameter as a function of TMV concentration. Aqueous (0.05M borate buffer, pH 8.5) solutions of viruses of 300 *nm* lengths. Solid line represents a hard-rod particles model, based on the Onsager theory.
(Reproduced with permission from ref.[8]).

magnetic field-induced birefringence (Δn_H), that was shown to be proportional to H^2. Due to the anisotropy of diamagnetic susceptibility of the TMV solutions, high magnetic field strengths are needed to induce measurable values of the induced birefringence. The maximum field employed in these experiments was $H \sim 13.5\ T$. Neutron scattering experiments reveal that, in the nematic mesophase of TMV solutions, the scattering peak observed perpendicularly to the phase director shows a Gaussian-type profile (Figure 9.3). Their results allowed an estimation of the number of viruses in a correlation volume, at the isotropic-to-nematic phase transition, being from 3 to 6 viruses [12]. Δn_H increases with the virus concentration in the solution, and decreases for increasing temperatures.

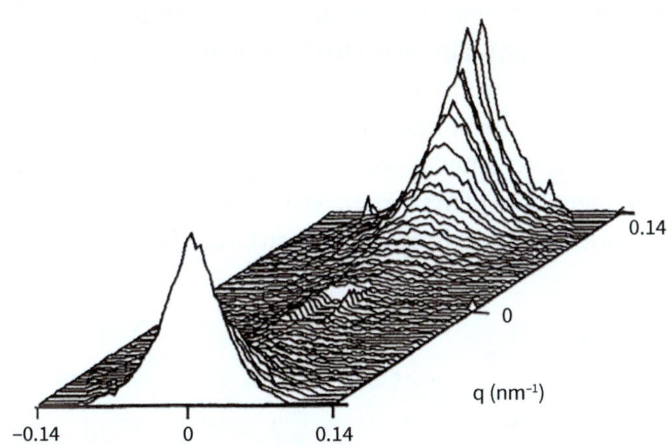

Figure 9.3 Small-angle neutron scattering from the mixture TMV/H_2O (low ionic strength), concentration of TMV 74 mg/mL, nematic phase magnetically oriented. *q* is the modulus of the scattering vector.
(Reproduced with permission from ref.[12]).

Freeze-fracture STEM (Scanning Transmission Electron Microscopy) images of TMV in the nematic phase revealed the sub-microscopic arrangement of the viruses, showing edge and screw disclinations in the structure, as theoretically predicted for nematics [13].

A smectic-A phase was observed in a 50 mM borate-buffer solution, pH of 8.5, and virus concentration in the range 170–190 mg/mL [14]. Samples were aligned in an external magnetic field of 2 T. Since the lengths of the virus and the smectic spacing are of the same order of magnitude, visible scattered light can be observed from smectic TMV samples illuminated with white light. The transition from nematic to smectic A was experimentally shown to be of second order, as theoretically predicted [12].

9.3 Filamentous phage virus (fd)

The fd viruses are long semiflexible rods, typical length of about 700 *nm*, diameter of 5 *nm*, and are chiral [15]. It is a polyelectrolyte, with charge density of about 15 e^{-}/nm. Figure 9.4 shows the fd virus, observed in electron microscopy [16].

Lapoint and Marvin [17] prepared an aqueous suspension of fd (ionic strength 0.01 M at $4.6 < pH < 12.0$), and observed a cholesteric mesophase (Ch). The cholesteric pitch observed is of the order of tens of micrometres. About 10^3 units (viruses) are present in a distance corresponding to a pitch.

Coexistence of isotropic and cholesteric phases was observed in different experimental conditions of virus concentration and ionic strength of the solution [12]. Mixture centrifugation is commonly used to separate the cholesteric phase from the isotropic, for further experiments in a single phase. Figure 9.5 shows the dependence of the cholesteric pitch as a function of virus concentration [12].

Solutions of fd viruses also stabilize the smectic mesophase (Figure 9.6). The viruses are stacked in layers with repeating distances of the order of magnitude of the wavelength of visible light. This leads to the brilliant iridescent aspect of the solution when illuminated with white light [18]. At a given ionic strength, by increasing the virus concentration in the buffer solution, the observed phase sequence is, typically, isotropic–cholesteric–smectic–disordered phase. The isotropic to cholesteric and cholesteric to smectic are of first order. An imposed twist in the smectic structure leads to the formation of regularly spaced Twist-Grain Boundaries (TGB) [12].

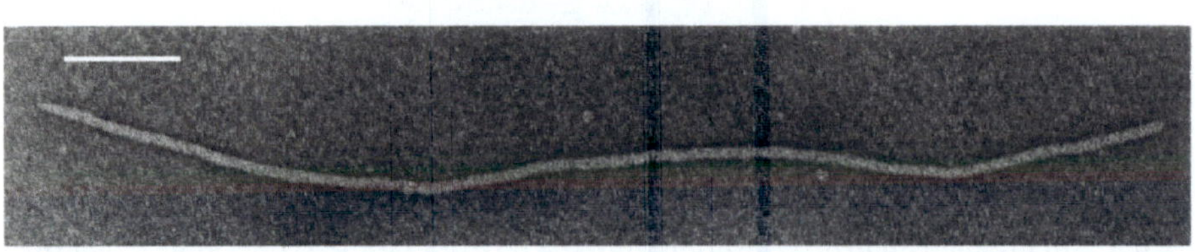

Figure 9.4 Electron microscopy image of the semi-flexible bacteriophages fd. The white bar corresponds to 0.1 μm.
(Reproduced with permission from ref.[16]).

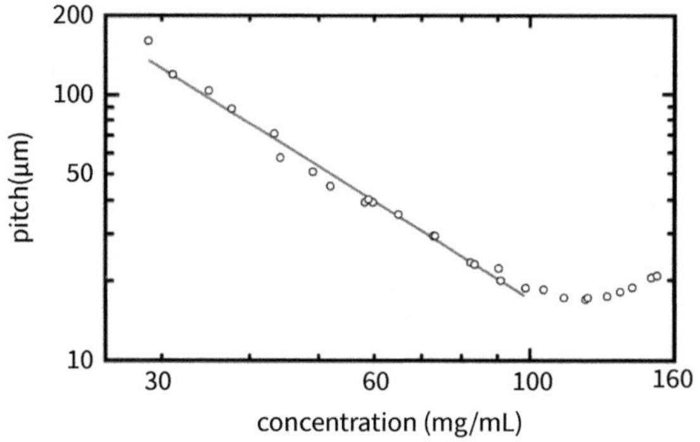

Figure 9.5 Suspension of fd virus in water (53 mM ionic strength buffer and pH 8), cholesteric phase. Dependence of the pitch on virus concentration.
(Reproduced with permission from ref.[12]).

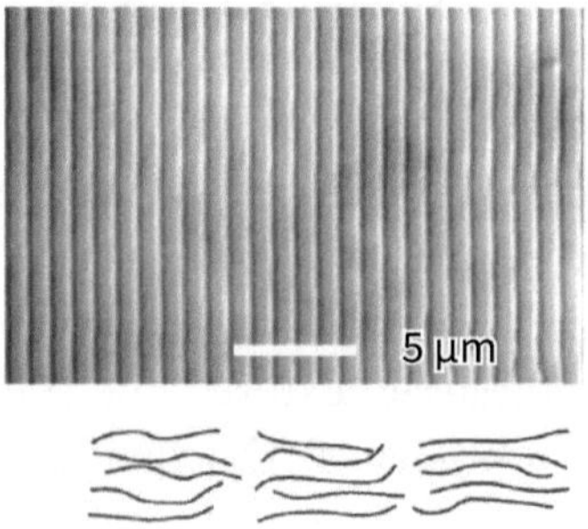

Figure 9.6 Differential interference contrast optical micrograph of fd solution in the smectic phase. The lines below the picture represent the oriented fd viruses.
(Reproduced with permission from ref.[18]).

Fd viruses can be 'decorated' with polymers, in aqueous suspensions. Polyethylene glycol (PEG) was grafted onto the viruses and, interestingly, the cholesteric phase was, still, stabilized [19]. The PEG coating was enough to maintain the individual viruses away from each other, at a distance of about 5 virus diameters. In these conditions, the electrostatic interaction between neighbour viruses is reduced with respect to the suspension without the PEG. The steric interaction between coated viruses is responsible for the helicoidal arrangement [20]. The cholesteric pitch formed was shown to depend on the ionic strength of the suspension.

9.4 DNA and RNA

The DNA molecule is constituted by two strands of nucleotides, structured in a right-handed double helix. Is a highly charged polyelectrolyte, with persistent length of about 50 nm [21]. In living cells, the length of the DNA macromolecule may achieve dimensions of the order of a metre. However, to show liquid crystalline behaviour, the length of this macromolecule must be reduced to dimensions of the order of tens of nanometres, with a diameter of about 2.2 nm. Suspensions of DNA in water, with ions (usually Na^+, K^+, and $NH_4{}^+$), may stabilize liquid crystalline mesophases, depending on the DNA concentration and temperature. The phase sequence, for increasing DNA concentration, is: isotropic–blue phase–cholesteric–columnar hexagonal–crystalline [21]. In the Onsager framework, suspensions of rod-shaped basic units (length L and diameter D) present liquid crystalline mesophases if $L/D > 4$, above a critical solute concentration. However, as we will discuss in this section, small DNA units, that individually do not fulfil these conditions, stabilize liquid crystalline phases, due to end-to-end adhesion of small units.

NMR experiments with aqueous suspensions of DNA were reported by Strzelecka and co-workers [22], identifying the cholesteric mesophase. Phase-coexistence regions were identified and the presence of a smectic phase was claimed (Figure 9.7) as a function of DNA concentration.

The cholesteric mesophase was extensively studied by Leforestier and Livolant [23] in DNA strands of 50 nm length (146 base pairs (bp)), dissolved in various solvents with concentrations ranging from 150 to 250 mg/mL. Polarized optical (POM) and electron microscopy were employed to observe textures and the structure of the cholesteric phase. Figure 9.8 shows a typical texture observed in POM of DNA dissolved in a TE (Tris-EDTA) buffer. Besides the helical pitch which is clearly identified in the cholesteric textures, the presence of defects (disclinations and dislocations) is also observed. Disclinations appear in regions of the

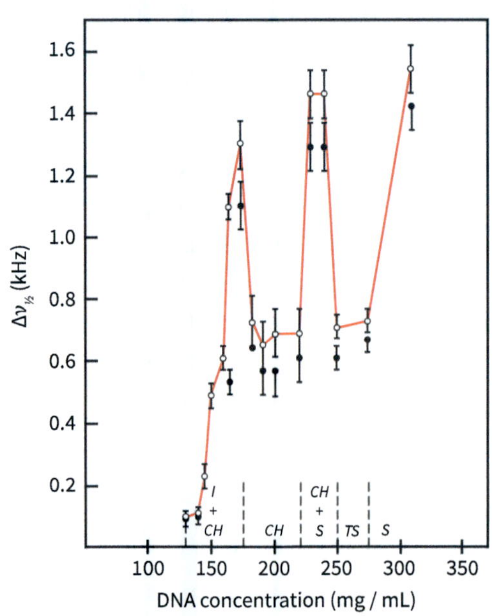

Figure 9.7 DNA/water experimental phase diagram: ^{31}P NMR linewidth as a function of DNA concentration. Open circles T = 30 °C; filled circles T = 50 °C. I, CH, and S represent the isotropic, cholesteric, and smectic phases.

Figure 9.8 Polarized optical microscopic texture of DNA dissolved in TE buffer (1mM EDTA in 10mM Tris-Cl-buffer, pH7.6) and 0.1M NaCl. Amplification × 847, black arrows indicate defects in the texture.
(Reproduced with permission from ref.[23]).

cholesteric structure where the director field is strongly bent. Dislocations are introduced or removed from the periodic local cholesteric structure, similar to lattices in crystals. With the help of the electron microscopy technique, where the sample is quickly frozen and a replica is made from a cut from the sample, dislocations and disclinations were identified (Figure 9.9). The pitch observed in these mixtures is, typically, of the order of μm. Interestingly, the behaviour of the cholesteric pitch was shown to be independent of the DNA concentration (Figure 9.10), at least in the range from 150 to about 280 mg/mL [24]. In some particular sample preparations (e.g., sonicated solutions), the pitch may reach values of the order of $\sim 0.3\,\mu m$ [23].

Between the isotropic and the cholesteric phases, blue phases were identified in DNA solutions. These phases (BP I, BP II, and BP III) were observed in suspensions with long DNA molecules (170–1700 nm long), sonicated for a long time [21, 24]. Blue phases were, mainly, identified by means of electron-microscopy freeze-fracture replicas [25]. Interestingly, and differing from the usual thermotropic blue phases, the pitch in DNA suspensions is much smaller than the one observed in the cholesteric mesophase of the same suspension. Since the length of the DNA molecule is much bigger than that of usual molecules in thermotropics,

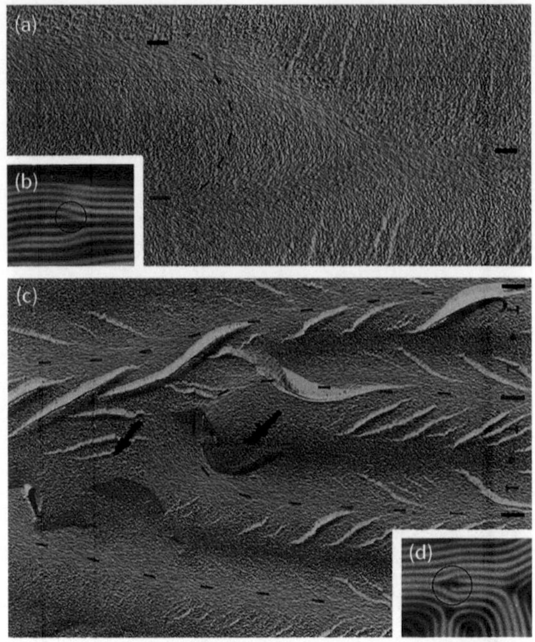

Figure 9.9 (a) and (c) Electron microscopy observations of edge dislocations in the cholesteric phase of DNA dissolved in 0.25 M ammonium acetate, 10 mM sodium cacodylate, and 0.5 mM EDTA (pH 7). (b) and (d) Corresponding POM textures of the edge dislocations. Magnification: (a) and (c) × 18, 750; (b) and (d) × 1, 050. Black arrows point to defects.
(Reproduced with permission from ref.[23]).

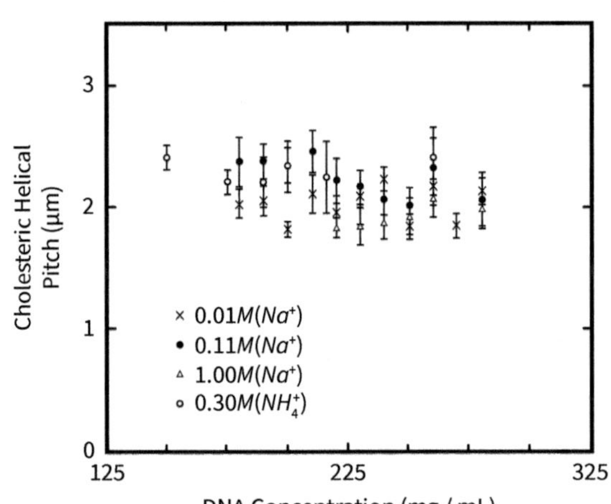

Figure 9.10 DNA molecules, 50 nm length, in a buffer (labelled in the lc DNA) in the presence of different cations. Cholesteric pitch as a function of the DNA concentration.

the structural details of the blue phases may be observed more easily in these suspensions, by using electron microscopy.

Increasing the DNA concentration, regardless of the length of the molecule (from 50 nm to some micrometres), a columnar phase is stabilized. Molecules form columns that are parallel to each other, arranged in a hexagonal structure, with the axes of the columns perpendicular to the plane where the 2D hexagonal arrangement is established [21] (Figure 9.11). The X-ray diffraction pattern of this mesophase clearly shows that it is not a crystalline phase, with bands that reveal the orientation of the columns with respect to the X-ray beam and the laboratory frame axes. The hexagonal lattice parameter observed is of the order of 3 nm. The columns are free to rotate around their cylindrical symmetry axis, and also may slide with respect to one another in the structure. The textures under POM are those characteristics of the hexagonal mesophase, with disclination lines strongly dependent on the particular sample composition and sample holder confinement geometry.

Short DNA strands (typically from 6 bp to 20 bp), in aqueous suspensions were shown to stabilize the chiral nematic (labelled N*, or cholesteric) mesophase, even though the shape anisotropy of the strands (i.e., length/diameter) is so small that a N* phase would not be expected [26]. The typical dimension of the DNA

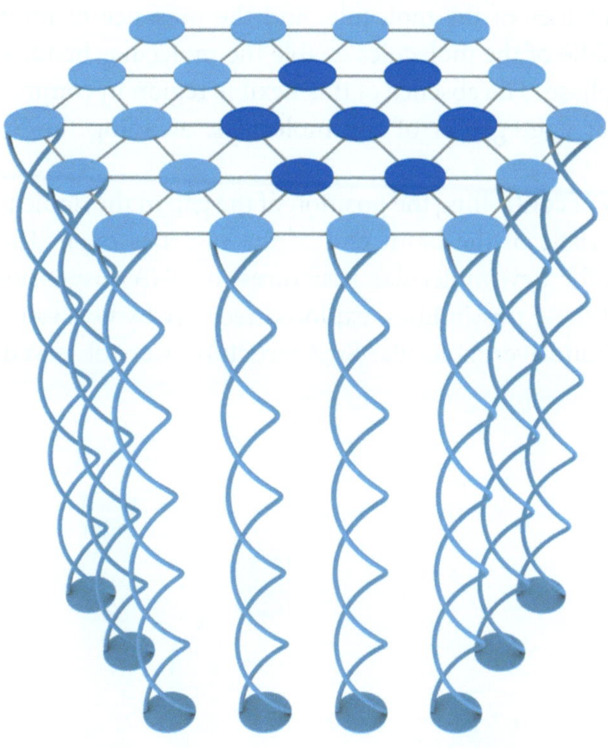

Figure 9.11 Sketch of the hexagonal columnar mesophase in suspensions of DNA. (Reproduced with permission from ref.[21]).

molecule corresponding to the bp described above ranges from about 2.0 to 6 nm. The mechanism proposed to increase the shape anisotropy of the basic units that stabilize the N* phase in the suspension, is the end-to-end adhesion of the small strands, and posterior nematic ordering of these aggregates (Figure 9.12). Hydrophobic/hydrophilic interactions at the ends of the short strands are essential to account for this adhesion [27]. Left- and right-handed helical arrangements were observed in the N* phase, with pitches ranging from values where the structure barely differs from that of the achiral nematic (i.e., macroscopic dimension), to about that of visible light [28]. The handedness of the structure was shown to be dependent on the length of the basic unit and the particular end-to-end interaction of them.

The stabilization of a smectic phase in a short DNA suspension was achieved with an ingenious trick of placing a flexible single-stranded DNA region in the middle of the duplex molecule [29]. Figure 9.13 shows a sketch of the full-paired duplex DNA and the gapped duplex. This gapped region, that corresponds to a single strand, introduces a flexibility to the entire structure, allowing the stabilization of a smectic A (SmA) mesophase in the suspension (Figure 9.13(c)). The SmA structure proposed was supported by small-angle X-ray scattering and optical polarized microscopy, as well as Monte Carlo simulations. The attractive interaction between the opened extremities of the molecule, and the presence of the gapped-flexible region in the middle of the molecules enable the molecular bending and stabilization of the SmA phase. The absence of this flexible region in common DNA strands favours the linear aggregation of the molecules, avoiding the SmA mesophase stabilization.

The possibility of controlling the position of the gap in the duplex DNA molecule and the characteristic of the blunt end of the moieties opened a broad field of research to look for new molecular structures in DNA suspensions. Figure 9.14 sketches some of these possibilities, explored recently by Gyawali and co-workers [30]. Bilayer or monolayer molecular SmA structures were observed.

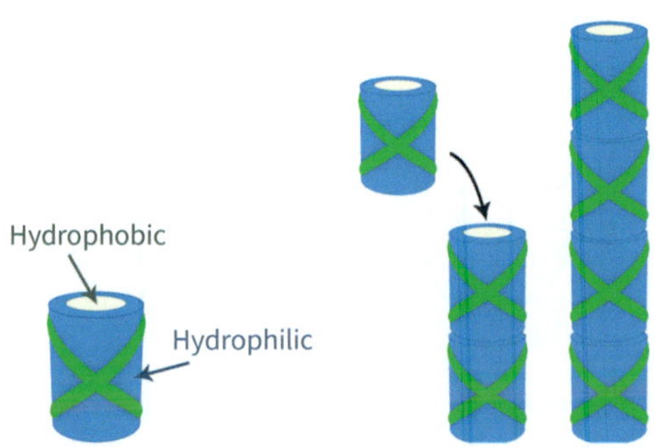

Figure 9.12 Sketch of the end-to-end adhesion of short DNA to form high anisometric basic units, to stabilize the nematic mesophase.

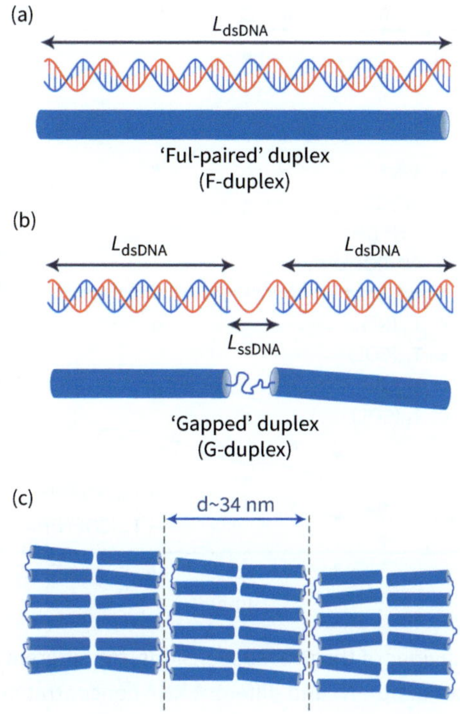

Figure 9.13 Sketch of (a) the full-paired duplex DNA and (b) the gapped duplex; (c) smectic A structure of the gapped-duplex DNA. L_{dsDNA}~ 16 nm.
(Reproduced with permission from ref.[29]).

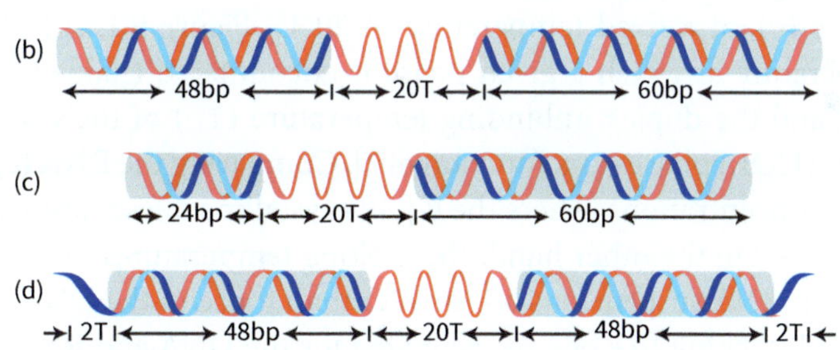

Figure 9.14 Schematic representations of gapped DNA molecular structures. nT and bp represent the number of thymine residues and base pairs, respectively. (a) symmetric; (b) and (c) asymmetric; (d) blunt ends with thymine.
(Reproduced with permission from ref.[30]).

In a similar way to what happens with DNA, suspensions of small RNA oligomers (from 6 to 20 bp) stabilize the cholesteric mesophase, with the helicoidal pitch ranging from ~260 nm to some micrometres, depending on the particular preparation [31]. As the RNA concentration is increased, the columnar mesophase is stabilized.

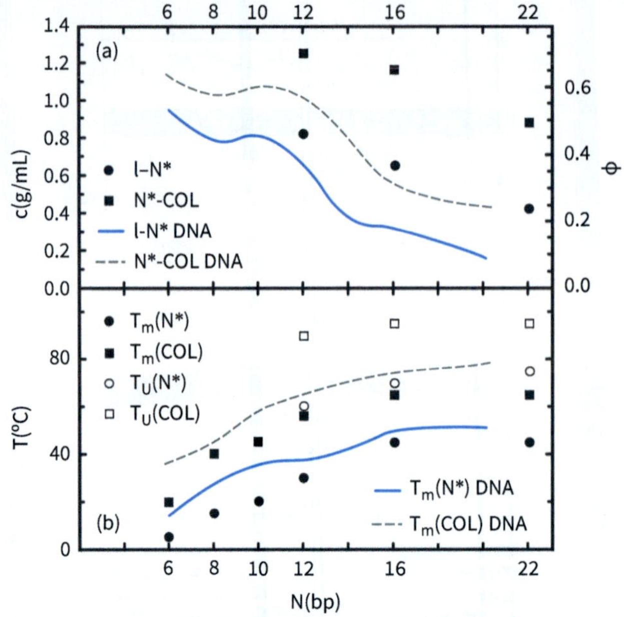

Figure 9.15 (a) Phase behaviour of the RNA suspensions, for different lengths of the oligomers (represented by the number of bp—N) and different RNA concentrations, c, and volume fraction ϕ at a fixed temperature of 20 °C.(b) Highest temperature at which a given phase is encountered (T_m) and the duplex unbinding temperature (T_U) of the cholesteric (N^*) and columnar (COL) phases, as a function of N.In both cases, DNA results are shown for comparison. (Reproduced with permission from ref.[31]).

Figure 9.15(a) shows the phase behaviour of the RNA suspensions, for different lengths of the oligomers (represented by the number of bp (N)) and different RNA concentrations (c), at a fixed temperature of 20 °C. Figure 9.15(b) illustrates the highest temperature at which a given phase is encountered (named melting temperature T_m) and the duplex unbinding temperature (T_U) of the cholesteric (N^*) and columnar (COL) phases as a function of N. Comparing the RNA (c_R) and DNA (c_D) oligomer concentrations, where the liquid crystalline phases are stabilized, one sees that $c_R > c_D$. On the other hand, the melting temperatures of the liquid crystalline phases in RNA suspensions are smaller than those in the DNA suspensions. The end-to-end stacking free energy evaluated for the DNA and RNA suspensions (at 20 °C) are about 3 kcal/mol (12.5kJ/mol) and 4 kcal/mol (16.7kJ/mol), respectively [31,32]. These energies in RNA suspensions lead to a significant temperature difference between T_U and T_m, with $T_U > T_m$.

The attractive interaction between duplexes at the ends of neighbour oligomers leads to pairing of nucleobases and energetically favours the end-to-end adhesion that stabilizes liquid crystal mesophases. Despite the difference between the small DNA and RNA end-to-end interaction and the face-to-face interaction between chromonic molecules, both processes lead to the formation of more anisometric basic units, essential to stabilizing the liquid crystalline phases.

9.5 Actin

Actin filament (F-actin) is a structural protein present in eukaryotic cells. It shows a filamentary twisted structure [33,34], with a helical pitch of about 40 nm and diameter of about 8 nm, as sketched in Figure 9.16. The persistent length of F-actin ranges from 6 μm to about 15 μm [35].

Rheological studies with F-actin isotropic solutions, typically at concentrations of about 20 mg/mL, shown shear-induced birefringence [36]. Liquid crystalline domains and thixotropy were observed in F-actin solutions under shear [37]. A nematic mesophase was observed in F-actin solutions [38], depending on the actin filament length and concentration (Figure 9.17). Strangely, no coexistence domains were reported in these experiments, despite the fact that the isotropic-to-nematic is a first-order phase transition. The measured order parameter in the nematic phase reached values of the order of 0.15, for filament lengths ranging from 2 μm to about 20 μm, at actin concentration around 4 mg/mL. The order parameter obtained is rather small, and increases monotonically with the filament length, which may indicate a weak first-order phase transition from the ordered to the isotropic phase. At a given concentration of actin the polydispersity of the F-actin macromolecules in solution affects the stabilization of the liquid crystalline mesophase. Flory and Abe treated this problem on a general basis and from the theoretical point of view [39].

More recently, Zhang and co-workers reported an interesting study of F-actin (about 7 nm in length) at the oil–water interface, stabilizing a nematic mesophase [40]. Thin films of the preparation allowed the investigation of defects in the nematic textures, and the topology of the director around these defects, and their dynamics (e.g., defect annihilation). Considering the Frank–Oseen elastic theory [41], where the nematic director topology is described in terms of three basic elastic distortions (splay, twist, and bend) with the corresponding elastic constants, it was possible to evaluate ratios between these elastic constants, depending on the particular sample preparation. Figure 9.18 shows the time evolution of the nematic texture, observed in optical microscopy, with the annihilation process of +1/2 and −1/2 defects.

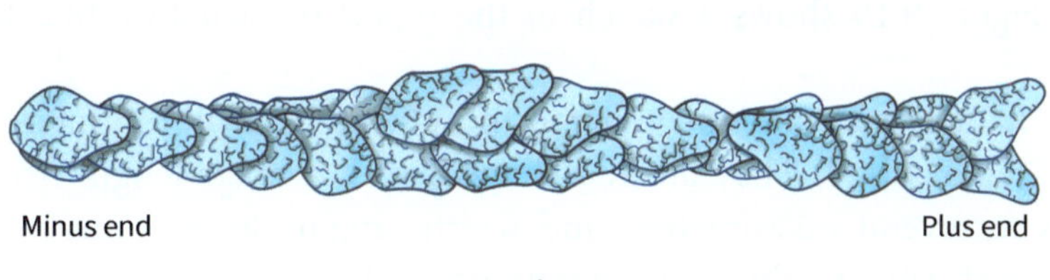

Figure 9.16 Sketch of an actin filament.

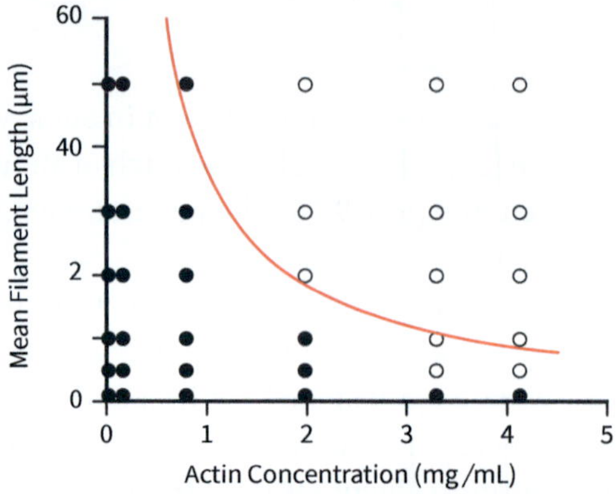

Figure 9.17 Phase diagram of the F-actin solution. Filled dots represent the isotropic solution and open dots the nematic liquid crystalline phase. Solid line represents the boundary between the anisotropic and isotropic regions, according to Flory's prediction for noninteracting rigid cylinders.
(Reproduced with permission from ref.[38]).

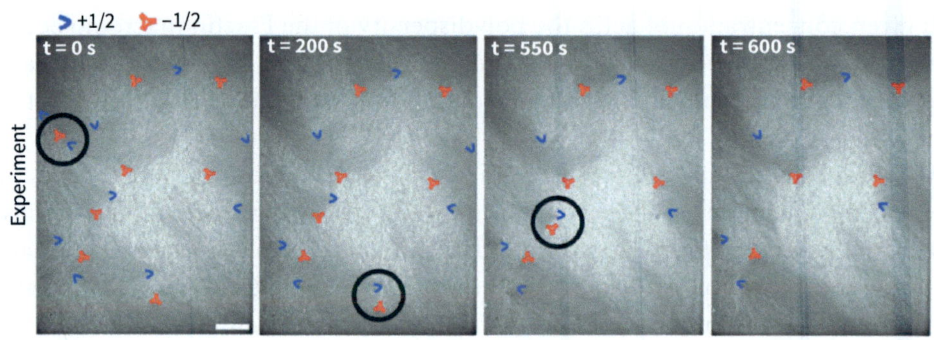

Figure 9.18 Consecutive optical microscopy textures of an F-actin preparation. Blue and red shapes represent +1/2 and −1/2 defects. Black circles indicate the defect annihilation events. White scale bar is 30 μm.
(Reproduced with permission from ref.[40]).

9.6 Cellulose nanocrystals and derivatives

Cellulose is a polysaccharide that forms a linear chain of β (1→4) linked D-glucose units [42]. Figure 9.19 shows a sketch of the repeating structure that forms the cellulose.

Cellulose nanocrystals are encountered in bulk cellulose fibres, intercalated by amorphous regions [43]. They are rod-shaped, with average length from tens of nanometres to several micrometres, and width ranging from 3 nm to about 50 nm. The aspect ratio of the nanocrystals depends on the source of the cellulose employed and the preparation method [44,45]. Nanocrystals are hydrophilic; however, their surface can be functionalized (e.g., esterification, etherification,

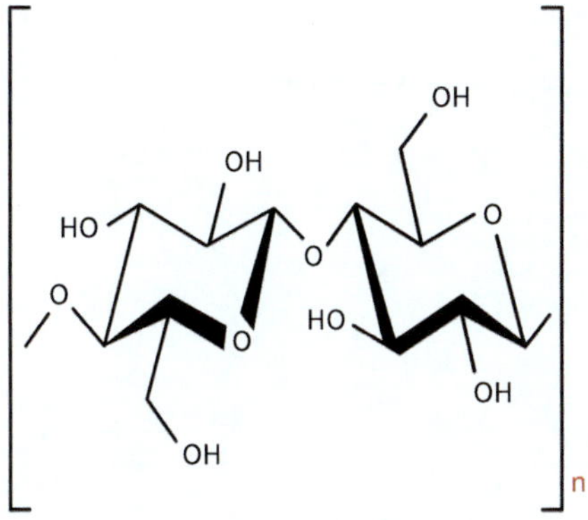

Figure 9.19 Sketch of the chemical structure of cellulose.

oxidation, amidation, carbamation, nucleophilic substitution, silylation, and polymer grafting) in order to be solubilized in nonpolar liquids and disperse in polymers.

Suspensions of cellulose nanocrystals in water, above a critical concentration of nanocrystals [45], stabilize the cholesteric lyotropic mesophase (Figure 9.20) with a pitch of the order of 30 μm [46]. The length of the nanocrystals ranges from 100 nm to 200 nm and the width is of about 5 nm. In the cholesteric structure, the long axes of the rod-shaped nanocrystals point perpendicularly to the direction of the helical axis [47], and the chirality is left-handed. As usual in the field of liquid crystals, shear was used to macroscopically align the rod-shaped particles. Below this critical concentration, which depends on the properties of the nanocrystals, the suspension is isotropic. As expected, a coexistence domain with both phases was identified between the isotropic and the cholesteric phase. The topology of the phase diagram of aqueous solutions of cellulose nanocrystals may be, coarsely, summarized as follows. The isotropic phase is stabilized in cellulose concentrations below about 3 vol %. Increasing the cellulose concentration to about 10 vol %, the cholesteric mesophase is stabilized.

An example of the shear-induced instabilities in aqueous suspensions of cellulose nanocrystals is shown in Figure 9.21 [48]. The solution is placed in between a glass blade and moving cover glass. The texture observed between crossed polarizers of an optical microscope shows the presence of hydrodynamic instabilities, formed with a periodicity along the shear imposed by the movement of the cover glass. Interestingly, by letting the solution in the cholesteric phase dry, films are formed, keeping the local helicoidal structure. This method allows the obtention of films with optical characteristics to be employed in different types of devices [49,50].

The interaction between individual cellulose nanocrystals, in solution, may be tuned by chemically modifying their surfaces (e.g., 2,2,6,6-tetramethylpiperidine-1-oxyl (TEMPO)-oxidized and Jeffamine polyetheramine M2070-grafted surface) [51]. These surface-decoration processes of the nanocrystals modify the temperature and critical concentration of cellulose to stabilize the cholesteric mesophase, due to the nature of the interparticle interaction forces after the surface modification.

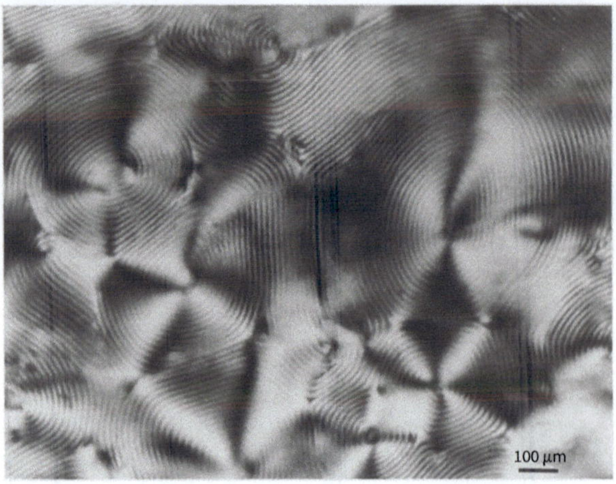

Figure 9.20 Texture of the cholesteric mesophase observed in polarized optical microscopy. Mixture of cellulose nanocrystals (about 3% by weight) and water.
(Reproduced with permission from ref.[46]).

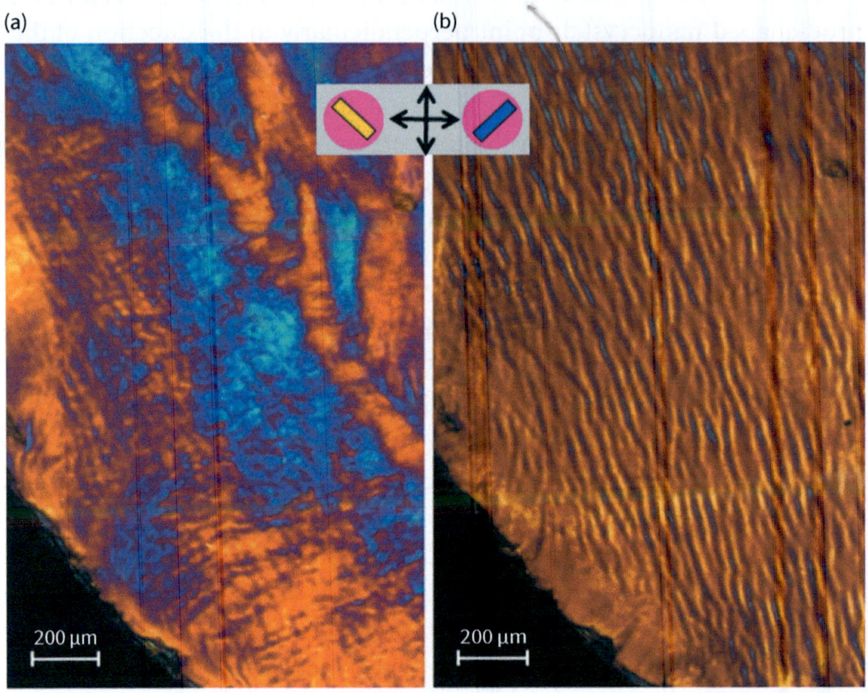

Figure 9.21 Textures of aqueous suspension of cellulose nanocrystals observed in polarized optical microscopy; 6 wt% carboxyl stabilized. Liquid placed between glass/cover glass blades: (a) freshly prepared; (b) after moving the cover glass from left to right, showing hydrodynamic instabilities formed, with lines perpendicular to the shear direction.
(Reproduced with permission from ref.[48]).

Cellulose derivatives are obtained by chemically modifying the original cellulose, by means of, e.g., esterification, etherification, and oxidation reactions of its hydroxyl groups [52]. Examples of these derivatives are: carboximethylcellulose (CMC), hydroxypropyl cellulose (HPC), cyanoethylcellulose (CEC), ethylcellulose (EC), methylcellulose (MC), hydroxyethylcellulose (HEC), hydroxypropylmethylcellulose (HPMC), and carboxymethylhydroxyethylcellulose (CMHEC).

Solutions of cellulose tricarbanilate (CTC) in methyl acrylate and methyl methacrylate stabilize the isotropic and nematic mesophases, as a function of the CTC concentration and temperature. The CTC concentration above which the nematic (or cholesteric, since the characterization is not completely clear) phase is stabilized is 33 wt % [53]. Solutions of HPC in water, in methanol, and in ethanol stabilize the cholesteric mesophase [54]. The minimum concentration of HPC necessary to form this phase in the solution is of the order of 40 wt %.

9.7 Living liquid crystals

The term 'living liquid crystals' was proposed by Zhou and co-workers [55] to designate a class of active matter composed of bacteria in a lyotropic, water-based, non-toxic liquid crystal. Interestingly, the individual and collective motions of the bacteria strongly couple with the short-range positional and long-range orientational ordering of the lyotropic liquid crystal.

Typical bacteria employed in experiments with liquid crystals show a uniaxial calamitic shape (cylindrical shape—length ~ $6\,\mu m$ and diameter ~ $0.5\,\mu m$), with flagella filaments attached to the main body. Rotation of the flagella pushes the bacterium's body forward.

Collective coherent large-scale movements of aerobic bacteria suspended in an isotropic fluid carrier were observed in 2004 [56]. The dynamic coherence of the collective movement occurs at high bacteria concentrations, where an individual interacts with others, essentially, by steric interactions. In vortices that can be formed, the collective velocity largely exceeds the typical individual velocity observed under dilute-solution conditions. Turbulent movement of drops of strains 1085 and YB886 of *Bacillus subtilis* in suspension were observed (Figure 9.22).

Sokolov and Aranson [57] investigated suspensions of strain 1085 of *Bacillus subtilis* in an isotropic fluid as a function of the amount of energy (via dissolved oxygen) provided to the bacteria, and their concentration. The physical mechanisms that rule the collective motion of the bacteria are of both short- (excluded-volume interaction in collisions) and long-range (hydrodynamic interactions) nature. The correlation length of collective motion was shown to be independent of the oxygen injection rate.

The great advantage of using liquid crystals in the preparation of suspensions with bacteria is the investigation of the spatial organization of them and their displacement, both individually and collectively, under different experimental conditions.

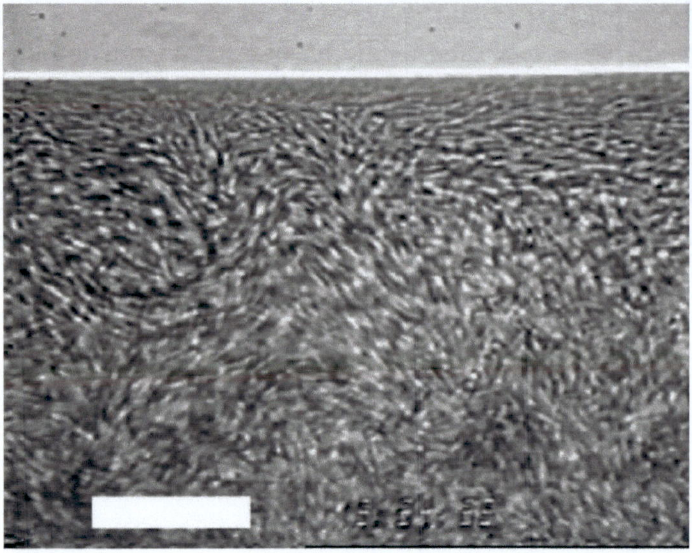

Figure 9.22 Turbulent motion of drops of strains 1085 and YB886 of *Bacillus subtilis* bacteria in an isotropic suspension. Optical microscope observation. The white bar represents 35 μm. (Reproduced with permission from ref.[56]).

For this purpose, the first aspect to be considered is the toxicity of the liquid crystal to different bacteria used in experimental assays. Woolverton and co-workers studied the toxicity of different liquid crystals (e.g., thermotropics, lyotropic chromonics, and surfactant-based lyotropics) to some bacteria [57]. The conclusion was that lyotropic chromonics did not show any toxicity to bacteria, making them ideal for the studies involving living liquid crystals [58].

A striking observation in living liquid crystals is that the trajectory of a single individual is affected by the local director configuration. Figure 9.23 shows an experiment where a single bacterium (*Bacillus subtilis*) travels in the nematic phase of a chromonic liquid crystal, oriented in planar geometry. At the temperature of the observation (e.g., by optical microscopy), the liquid crystal presents a phase coexistence of the nematic and isotropic phases. In the figure, the black region shows an isotropic phase, surrounded by the nematic phase, where the director is distorted at the interface. The bacterium follows this distorted-director region as time goes on [55]. After following this curved trajectory, the bacterium swims parallel to the nematic planarly oriented director field. Another spectacular observation is that the bacterium's flagella (typically a helical arrangement of filaments in a cylindrical geometry of diameter $\sim 25\, nm$), is able to distort the previously planar alignment of the director in scales much larger than its dimension ($\sim \mu m$).

The collective behaviour of swimming bacteria suspended in a planarly aligned nematic chromonic liquid crystal provokes distortions of the nematic director orientation in the bulk (Figure 9.24). Disclinations of charge $+1/2$ and $-1/2$ are formed

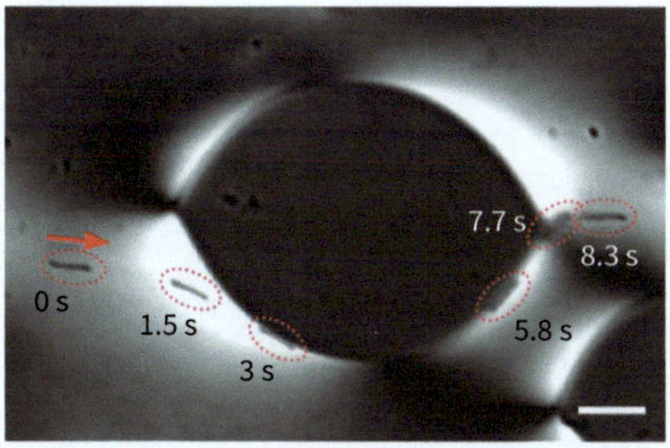

Figure 9.23 Time evolution of the *Bacillus subtilis* position in a chromonic liquid crystal, around an isotropic droplet in a nematic mesophase (coexistence between nematic and isotropic phases). Polarized optical microscope observation. White bar corresponds to 5μm.
(Reproduced with permission from ref.[55]).

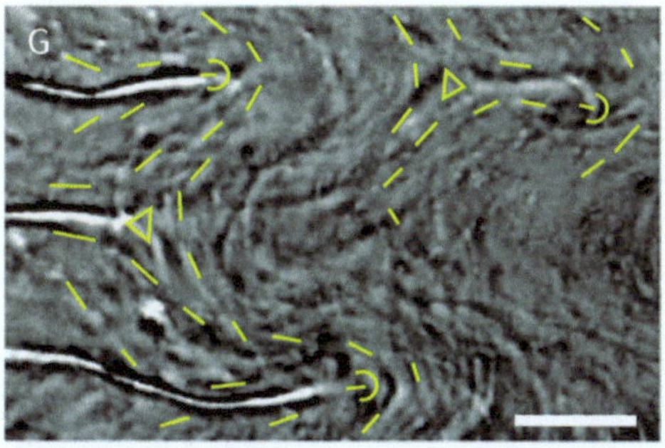

Figure 9.24 *Bacillus subtilis* bacteria suspended in a planarly oriented chromonic liquid crystal in the nematic phase. Semicircles and triangles represent disclinations of strength +1/2 and −1/2, respectively. White bar corresponds to 100 μm.
(Reproduced with permission from ref.[55]).

and annihilate as a function of time (Figure 9.25). A characteristic length scale (ξ) is present in the textures observed in Figures 9.24 and 9.25, corresponding to values of the order of $10^2 \mu m$ in those experiments. It was shown that ξ depends on the bacteria concentration, the elastic constant of the liquid crystal, and the sample thickness.

From the theoretical point of view, Genkin and co-workers [59,60] developed a model where two populations of bacteria travel in opposite directions (Figure 9.26). The bacteria swimming in opposite directions allows the formation of convergent and depleted regions of individuals, resulting in the topological defects $+1/2$ and $-1/2$, respectively.

As a final remark, biological basic units like those discussed in this chapter, dispersed in a liquid carrier, constitutes a field of research with implications not only in fundamental physical chemistry, but also in technological applications, with a broad interface with biology and medicine. Drug delivery is only one of these applications; however, liquid crystal properties seem to have deep and fundamental implications in the maintenance and evolution of life as we know it.

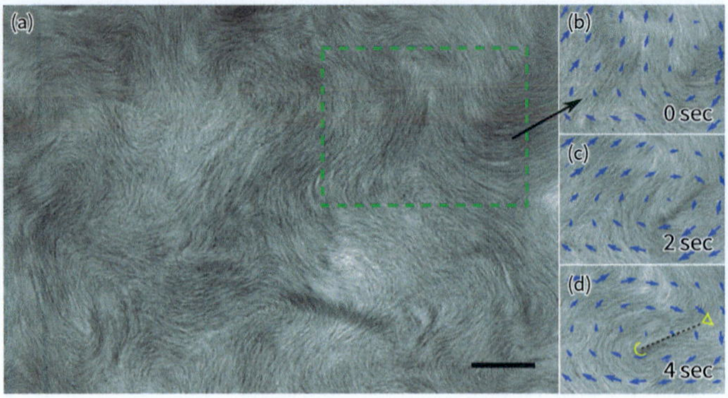

Figure 9.25 *Bacillus subtilis* bacteria suspended in a planarly oriented chromonic liquid crystal in the nematic phase. Defects are generated by the swimming bacteria, suspended in the nematic phase. (a) Texture observed with many disclinations; (b), (c), and (d) evolution of the region in the green square as a function of time. Blue arrows represent the direction of the bacteria movement. Semicircles and triangles represent disclinations of strength +1/2 and −1/2, respectively. White bar corresponds to 30 μm.
(Reproduced with permission from ref.[55]).

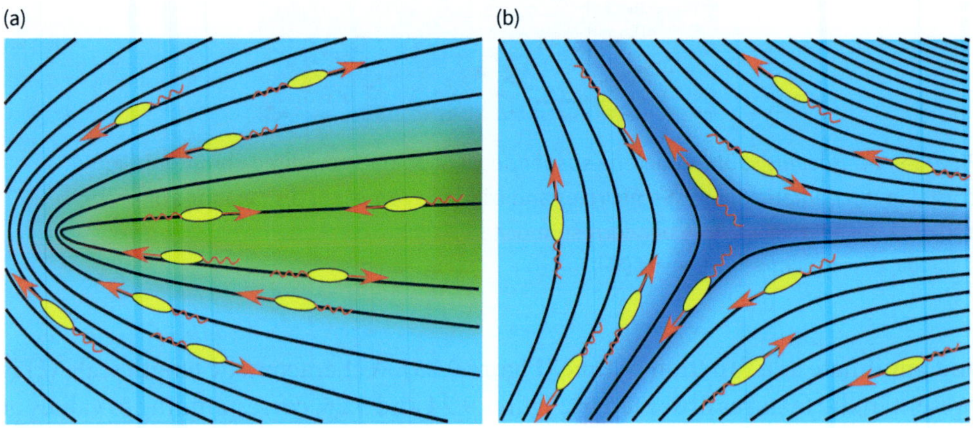

Figure 9.26 Representation of (a) accumulation and (b) depletion of bacteria in the vicinity of +1/2 and −1/2 defects, respectively.
(Reproduced with permission from ref.[59]).

References

[1] G. H. Brown and J. J. Wolken (1979). *Liquid Crystals and Biological Structures.* New York: Academic Press.

[2] E. V. Orlova, M. B. Sherman, W. Chiu, H. Mowri, L. C. Smith, and A. M. Gotto, Jr. (1999). Three-dimensional structure of low density lipoproteins by electron cryomicroscopy. *PNAS*, 96, 8420–8425.

[3] T. Hevonoja, M. O. Pentikäinen, M. T. Hyvönen, P. T. Kovanen, and M. Ala-Korpela (2000). Structure of low density lipoprotein (LDL) particles: Basis for understanding molecular changes in modified LDL. *Biochimica et Biophysica Acta*, 1488, 189–210.

[4] H. Lecoq (2001). Discovery of the first virus, the tobacco mosaic virus: 1892 or 1898? *C. R. Acad. Sci. III*, 324, 929–933.

[5] S. Sanctis, R. C. Hoffmann, S. Eiben, and J. J. Schneider (2015). Microwave assisted synthesis and characterisation of a zinc oxide/tobacco mosaic virus hybrid material. An active hybrid semiconductor in a field-effect transistor device. *Beilstein J. Nanotechnol.*, 6, 785–791.

[6] A. Klug and D. L. D. Caspar (1961). The structure of small viruses. *Adv. in Virus Research*, 7, 225–325.

[7] H. Graf and H. Löwen (1999). Phase diagram of tobacco mosaic virus solutions. *Phys. Rev. E*, 59, 1932–1942.

[8] R. Oldenbourg, X. Wen, R. B. Meyer, and D. L. D. Caspar (1988). Orientational distribution function in nematic tobacco-mosaic-virus liquid crystals measured by X-ray diffraction. *Phys. Rev. Lett.*, 61, 1851–1854.

[9] P. Sheng (1976). Hard rod model of the nematic-isotropic phase transition. In: E. B. Priestley, P. J. Wojtowicz, and P. Sheng (eds), *Introduction to Liquid Crystals*, New York: Plenum Press, pp 59–70.

[10] S. Fraden, G. Maret, D. L. D. Caspar, and R. B. Meyer (1989). Isotropic-nematic phase transition and angular correlations in isotropic suspensions of tobacco mosaic virus. *Phys. Rev. Lett.*, 63, 2068–2071.

[11] S. Fraden, G. Maret, and D. L. D. Caspar (1993). Angular correlations and the isotropic-nematic phase transition in suspensions of tobacco mosaic virus. *Phys. Rev. E*, 48, 2816–2837.

[12] S. Fraden (1995). In: M. Baus, L. F. Rull, and J.-P. Ryekaert (eds), *Observation, Prediction and Simulation of Phase Transitions in Complex Fluids*, Dordrecht: Springer Science+Business Media, B.V.

[13] J. A. N. Zasadzinski, M. J. Sammon, R. B. Meyer, M. Cahoon, and D. L. D. Caspar (1986). Freeze-fracture imaging of ordered phases of tobacco mosaic virus in water. *Mol. Cryst. Liq. Cryst.*, 138, 211–229.

[14] X. Wen, R. B. Meyer, and D. L. D. Caspar (1989). Observation of smectic-A ordering in a solution of rigid-rod-like particles. *Phys. Rev. Lett.*, 63, 2760–2763.

[15] D. A. Marvin and H. Hoffmann-Berling (1963). Physical and chemical properties of two new small bacteriophages. *Nature*, 197, 517–518.

[16] C. W. Gray, R. S. Brown, and D. A. Marvin (1981). Adsorption complex of filamentous fd virus. *J. Mol. Biol.*, 146, 621–627.

[17] J. Lapointe and D. A. Marvin (1973). Filamentous bacterial viruses VIII. Liquid crystals of fd. *Mol. Cryst. Liq. Cryst.*, 19, 269–278.

[18] Z. Dogic and S. Fraden (1997). Smectic phase in a colloidal suspension of semiflexible virus particles. *Phys Rev. Lett.*, 78, 2417–2420.

[19] E. Grelet and S. Fraden (2003). What is the origin of chirality in the cholesteric phase of virus suspensions? *Phys. Rev. Lett.*, 90, 198302.

[20] Z. Dogic and S. Fraden (2006). Ordered phases of filamentous viruses. *Current Opinion in Colloid & Interface Science*, 11, 47–55.

[21] F. Livolant and A. Leforestier (1996). Condensed phases of DNA: Structures and phase transitions. *Prog. Polym. Sci.*, 21, 1115–1164.

[22] T. E. Strzelecka, M. W. Davidson, and R. L. Rill (1988). Multiple liquid crystal phases of DNA at high concentrations. *Nature*, 331, 457–460.

[23] F. Livolant and A. Leforestier (1993). Supramolecular ordering of DNA in the cholesteric liquid crystalline phase: an ultrastructural study. *Biophys. J.*, 65, 56–72.

[24] D. H. Van Winkle, M. W. Davidson, W. Chen, and R. L. Rill (1990). Cholesteric helical pitch of near persistence length DNA. *Macromolecules*, 23, 4140–4148.

[25] A. Leforestier and F. Livolant (1994). DNA liquid crystalline blue phases. Electron microscopy evidence and biological implications. *Liquid Crystals*, 17, 651–658.

[26] M. Nakata, G. Zanchetta, B. D. Chapman, C. D. Jones, J. O. Cross, R. Pindak, T. Bellini, and N. A. Clark (2007). End-to-end stacking and liquid crystal condensation of 6- to 20-base pair DNA duplexes. *Science*, 318, 1276–1279.

[27] G. Zanchetta (2009). Spontaneous self-assembly of nucleic acids: liquid crystal condensation of complementary sequences in mixtures of DNA and RNA oligomers. *Liquid Crystals Today*, 18, 40–49.

[28] G. Zanchetta, F. Giavazzi, M. Nakata, M. Buscaglia, R. Cerbino, N. A. Clark, and T. Bellini (2010). Right-handed double-helix ultrashort DNA yields chiral nematic phases with both right- and left-handed director twist. *PNAS*, 107, 17497–17502.

[29] M. Salamonczyk, J. Zhang, G. Portale, C. Zhu, E. Kentzinger, J. T. Gleeson, A. Jakli, C. De Michele, J. K. G. Dhont, S. Sprunt, and E. Stiakakis (2016). Smectic phase in suspensions of gapped DNA duplexes. *Nat. Commun.*, 7, 13358.

[30] P. Gyawali, R. Saha, G. P. Smith, M. Salamonczyk, P. Kharel, S. Basu, R. Li, M. Fukuto, J. T. Gleeson, N. A. Clark, A. Jákli, H. Balci, and S. Sprunt (2021). Mono- and bilayer smectic liquid crystal ordering in dense solutions of 'gapped' DNA duplexes. *PNAS*, 118, e2019996118.

[31] G. Zanchetta, T. Bellini, M. Nakata, and N. A. Clark (2008). Physical polymerization and liquid crystallization of RNA oligomers. *J. Am. Chem. Soc.*, 130, 12864–12865.

[32] G. Zanchetta, M. Nakata, M. Buscaglia, N. A. Clark, and T. Bellini (2008). Liquid crystal ordering of DNA and RNA oligomers with partially overlapping sequences. *J. Phys.: Condens. Matter*, 20, 494214.

[33] E. H. Egelman (1985). The structure of F-actin. *J. Muscle Research and Cell Motility*, 6, 129–151.

[34] E. H. Egelman, N. Francist, and D. J. DeRosiert (1982). F-actin is a helix with a random variable twist. *Nature*, 298, 131–135.

[35] R. Furukawa, R. Kundra, and M. Fechheimer (1993). Formation of liquid crystals from actin filaments. *Biochemistry*, 32, 12346–12352.

[36] R. E. Buxbaum, T. Dennerll, S. Weiss, and S. R. Heidemann (1987). F-actin and microtubule suspensions as indeterminate fluids. *Science*, 235, 1511–1514.

[37] A. Kerst, C. Chmielewski, C. Livesay, R. E. Buxbaum, and S. R. Heidemann (1990). Liquid crystal domains and thixotropy of filamentous actin suspensions. *PNAS*, 87, 4241–4245.

[38] C. M. Coppin and P. C. Leavis (1992). Quantitation of liquid-crystalline ordering in F-actin solutions. *Biophys. J.*, 63, 794–807.

[39] P. J. Flory and A. Abe (1978). Statistical thermodynamics of mixtures of rodlike particles. 1. Theory for polydisperse systems. *Macromolecules*, 11, 1119–1122.

[40] R. Zhang, N. Kumar, J. L. Ross, M. L. Gardel, and J. J. de Pablo (2017). Interplay of structure, elasticity, and dynamics in actin-based nematic materials. *PNAS*, 115, E124–E133.

[41] F. C. Frank (1958). Liquid crystals. On the theory of liquid crystals. *Discussions of the Faraday Society*, 25, 19–28.

[42] Y. Nishiyama, P. Langan, and H. Chanzy (2002). Crystal structure and hydrogen-bonding system in cellulose Iβ from synchrotron X-ray and neutron fiber diffraction *J. Am. Chem. Soc.*, 124, 9074–9082.

[43] R. H. Newman and J. A. Hemmingson (1995). Carbon-13 NMR distinction between categories of molecular order and disorder in cellulose. *Cellulose*, 2, 95–110.

[44] J. George and S. N. Sabapathi (2015). Cellulose nanocrystals: synthesis, functional properties, and applications. *Nanotechnology, Science and Applications*, 8, 45–54.

[45] R. J. Moon, A. Martini, J. Nairn, J. Simonsen, and J. Youngblood (2011). Cellulose nanomaterials review: structure, properties and nanocomposites. *Chem. Soc. Rev.*, 40, 3941–3994.

[46] J. F. Revol, H. Bradford, J. Giasson, R. H. Marchessault, and D. G. Gray (1992). Helicoidal self-ordering of cellulose microfibrils in aqueous suspension. *Int. J. Biol. Macromol.*,14, 170–172.

[47] D. G. Gray and X. Mu (2015). Chiral nematic structure of cellulose nanocrystal suspensions and films; polarized light and atomic force microscopy. *Materials*, 8, 7873–7888.

[48] D. G. Gray (2016). Recent advances in chiral nematic structure and iridescent color of cellulose nanocrystal films. *Nanomaterials*, 6, 213.

[49] J. P. F. Lagerwall, C. Schütz, M. Salajkova, J. Noh, J. H. Park, G. Scalia, L. Bergström (2014). Cellulose nanocrystal-based materials: from liquid crystal self-assembly and glass formation to multifunctional thin films. *NPG Asia Mater.*, 6, e80.

[50] M. H. Godinho, J. G. Fonseca, A. C. Ribeiro, L. V. Melo, and P. Brogueira (2002). Atomic force microscopy study of hydroxypropylcellulose films prepared from liquid crystalline aqueous solutions. *Macromolecules*, 35, 5932–5236.

[51] F. Azzam, L. Heux, and B. Jean (2016). Adjustment of the chiral nematic phase properties of cellulose nanocrystals by polymer grafting. *Langmuir*, 32, 4305–4312.

[52] R. Khiari and M. N. Belgacem (2017). Potential for using multiscale *Posidonia oceanica* waste: Current status and prospects in material science. In: M. Jawaid, P. M. Tahir, and N. Saba (eds), *Lignocellulosic Fibre and Biomass-Based Composite Materials, Processing, Properties and Applications*, Cambridge: Woodhead Publishing, pp 447–471. doi.org/10.1016/C2015-0-04050-3

[53] J. M. G. Cowie, V. Arrighi, J. Cameron, I. McEwan, and I. J. McEwen (2001). Lyotropic liquid crystalline cellulose derivatives in blends and molecular composites. *Polymer*, 42, 9657–9663.

[54] R. S. Werbowyj and D. G. Gray (1980). Ordered phase formation in concentrated hydroxpropylcellulose solutions. *Macromolecules*, 13, 69–73.

[55] S. Zhou, A. Sokolov, O. D. Lavrentovich, and I. S. Aranson (2014). Living liquid crystals. *PNAS*, 111, 1265–1270.

[56] C. Dombrowski, L. Cisneros, S. Chatkaew, R. E. Goldstein, and J. O. Kessler (2004). Self-concentration and large-scale coherence in bacterial dynamics. *Phys Rev. Lett.*, 93, 098103.

[57] A. Sokolov and I. S. Aranson (2012). Physical properties of collective motion in suspensions of bacteria. *Phys. Rev. Lett.*, 109, 248109.

[58] C. J. Woolverton, E. Gustely, L. Li, and O. D. Lavrentovich (2005). Liquid crystal effects on bacterial viability. *Liquid Crystals*, 32, 417–423.

[59] M. M. Genkin, A. Sokolov, O. D. Lavrentovich, and I. S. Aranson (2017). Topological defects in a living nematic ensnare swimming bacteria. *Phys. Rev. X*, 7, 011029.

[60] M. M. Genkin, A. Sokolov, and I. S. Aranson (2018). Spontaneous topological charging of tactoids in a living nematic. *New J. Phys.*, 20, 043027.

10
Rods, tubes, wires, and sheets

Lyotropic liquid crystals formed by anisotropic nanomaterials is obviously a topic with diffuse borders and overlaps with other chapters of this book, for example that of inorganic and mineral liquid crystals as we have seen from the nanotubes of imogolite. Similarly, one could, of course, ask where does a nanorod end and a nanowire start? We presume that one could introduce a rather arbitrary criterion, calling nanorods materials with an aspect ratio of 10 or below, while a nanowire would be a material with a very much larger aspect ratio. A number of the materials discussed below will effectively overlap with the discussion of Chapter 8 on inorganic liquid crystals. While in the latter chapter we have introduced mainly materials found in nature, we will here summarize findings on mainly manmade nanomaterials or organic and biological nanomaterials.

Talking about nanomaterials in relation to liquid crystals, one will find that the most dominant amount of literature is in fact related to dispersions of nanomaterials, nanoparticles, nanotubes, etc., in thermotropic liquid crystals. There are several reasons for adding nanoparticles to LCs, the first one being to tune the liquid crystal properties for applications, often displays, but also a whole range of other applications in nanotechnology have been proposed [1]. A whole range of material parameters can be affected by the addition of nanoparticles, the most important ones being the threshold voltage V_{th}, the birefringence Δn, and the dielectric anisotropy $\Delta\varepsilon$, but also viscosities and elastic constants. Particles are often of the order of 5–200 nm, for example zinc oxide ZnO, titanium dioxide TiO_2, aluminium oxide Al_2O_3, or barium titanate $BaTiO_3$. These materials are in fact bulk commodities, being used in cosmetics, paints, and as fillers in brick and plastics. The non-centrosymmetric structures of barium titanate, which generally depend on production and nanoparticle size, are ferroelectric.

To give some examples of liquid crystal–nanoparticle dispersions and their influences on material properties, we may commence with ZnO in nematics [2–7]. Manohar et al. [2] studied the electrooptic and dielectric properties of ZnO doped 5CB and found the dielectric anisotropy to slightly decrease for the addition of ZnO, while the threshold voltage nearly doubled. Nevertheless, the addition of the nanoparticles reduced the overall switching time (the sum of switching on and switching off) significantly, while both elastic splay constant and rotational viscosity increased—the latter quite drastically. These trends are in fact strongly dependent on the size and production of the nanoparticles and may vary. Mishra et al. [3] varied the shape of the zinc oxide from nanospheres to nanorods and nanowires and found that increasing and decreasing trends of physical quantities of the thermotropic LC

Lyotropic Liquid Crystals. Ingo Dierking and Antônio Martins Figueiredo Neto, Oxford University Press.
 DOI: 10.1093/9780191924927.003.0010

composite change with particle geometry. Other investigations were concerned with the concentration dependence of the electrooptic properties of ZnO nanoparticles [4], the changes in birefringence [5], thermodynamic properties [6] and the switching behaviour [7]. Similar studies were also carried out on other thermotropic liquid crystalline phases, such as ferroelectric SmC* materials [8,9].

Another popular nano-dopant in thermotropic liquid crystals is titanium dioxide, TiO_2. Pathak et al. [10] presented a detailed investigation of low, medium, and high birefringent materials varying the dopant concentration. Dielectric and electrooptic studies by Huang et al. [11] showed a decreasing threshold voltage as a function of TiO_2 concentration, which led to a decrease in elastic splay constant K_{11} and rotational viscosity. These trends were confirmed in a publication by Ayeb et al. [12]. Further studies of TiO_2 nanoparticle-doped nematics often include reference to applications, such as photoluminescence and threshold studies for displays and photovoltaic devices [13], the electrooptics of waveguides [14], an improvement of the dynamic response of liquid crystal lenses [15], doped polymer dispersed and polymer-stabilized liquid crystals for smart films [16], or low-power operation of vertically aligned nematic (VAN) cells [17]. Of course also other phases were investigated in relation to dispersed TiO_2 nanoparticles, for instance the cholesteric phase [18] and ferroelectric liquid crystals (FLCs) [19].

Some studies have also been carried out on aluminium oxide, for example by Yadav et al. [20], showing a decrease in conductivity, a decrease in permittivity and dielectric anisotropy, and a decrease of the relaxation frequency of the molecular flip-reorientation around the short molecular axis, for increasing nanoparticle concentration. The electrooptic response time, τ_{off}, on the other hand, was found to decrease with increasing nanoparticle concentration, which is in general beneficial for applications. Some rather diverse investigations have also been carried out in the past, for example the doping of bent-core liquid crystals [21] and rheological studies of nanoparticle doped nematics [22].

As mentioned before, barium titanate, $BaTiO_3$, has ferroelectric properties once the nanoparticles exceed a size of approximately 100 nm. For smaller sizes the particles are paraelectric. But it should be mentioned that the dispersion properties also depend strongly on the preparation methods [23], not only of the dispersion, but also on the preparation of the nanoparticles themselves. Similar to the above-mentioned studies, for $BaTiO_3$ there have also been investigations of transitions and electrooptics as a function of particle size [24]. Asymmetric responses of the Freedricksz transition were reported in relation to the application of positive versus negative electric fields [25]. A detailed study found that the birefringence increases by doping, while the threshold decreases, leading also to faster response times [26]. A range of reports include high-resolution calorimetric studies, where it is found that the transition temperature decreases for increasing nanoparticle concentration, while the transition gets increasingly smeared out [27]. An interesting soft memory effect for barium titanate–LC dispersions was shown by Basu [28], and a photorefractive cell has been demonstrated [29]. In the THz region the doped nanoparticles cause an increase of the birefringence by about 10% [30], enhanced conductivity is observed for dispersions with discotic materials in the columnar phase [31], and scattering is increased for polymer-dispersed liquid crystal (PDLC) scattering devices [32]. Al-Zangana et al. [33] published a comprehensive investigation of a comparison of size-dependent paraelectric and ferroelectric $BaTiO_3$ on the properties of the nematic and the ferroelectric SmC* phase. A recent review with a more comprehensive list of references was published by Garbovskiy and Glushchenko [34], discussing recent progress and future challenges in liquid crystal dispersions containing ferroelectric nanoparticles.

10.1 Nanorods

10.1.1 Tobacco Mosaic Virus, TMV

After that short overview of nanoparticles in thermotropic liquid crystals, it is now time to come back to the discussion of lyotropic systems. Besides rod-like, inorganic and mineral nanoparticles the nanorods of the tobacco mosaic virus (TMV) are one of the oldest recognized systems to form lyotropic phases. We will thus commence with this biological colloid. This was already briefly discussed in the previous chapter, especially with its relation to biological matter, and we here would like to place an emphasis on its colloidal properties in relation to liquid crystalline behaviour. TMVs are single-stranded RNA viruses which form a right-handed helical structure as shown in Figure 10.1(a). The virus affects a large range of plants, predominantly tobacco plants, where the infected leaves exhibit a mosaic-like pattern, hence the name. The shape of the TMV is very much rod-like and monodisperse with a length of approximately 200–300 nm and a width of about 20 nm (Figure 10.1(b)).

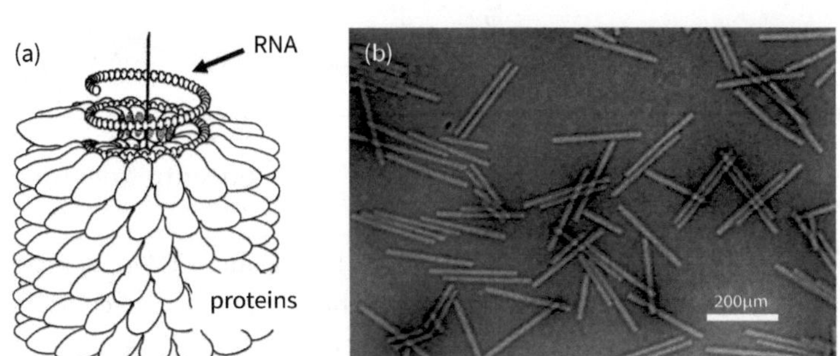

Figure 10.1 (a) Structure of the tobacco mosaic virus according to the last X-ray investigations of Rosalind Franklin, later published in a review by one of her co-workers [35]. (b) SEM image of tobacco mosaic viruses displaying a very monodisperse length and width distribution and an aspect ratio of approximately 15.

(Figure (a) reproduced with permission from ref.[35]).

The first investigations of the lyotropic liquid crystalline behaviour of the tobacco mosaic virus go back to Bawden and co-workers [36] in 1936 who describe the birefringence and tactoids of a TMV nematic phase between crossed polarizers, after, one year before, Stanley had reported the isolation of the crystalline protein of the TMV as small needles [37]. The size, shape, and physical characteristics such as diffusion and viscosity were determined by Lauffer [38] in 1944, before a more pronounced interest in the liquid crystalline behaviour of TMVs returned with the magnetic field-oriented X-ray investigations of Oldenbourg et al. [39] more than four decades later. These investigations provided a much-enhanced understanding of the LC phase by determining the orientational order distribution function and the order parameter of the nematic lyotropic phase. In fact, it was shown that this followed closely the predictions of Onsager's theory, making the TMVs an experimental model for hard-rod systems. At low volume fractions of TMVs in water an isotropic, diluted suspension is observed, which stays isotropic for increasing TMV concentration until a critical concentration is reached and the system enters a biphasic region. Passing the biphasic regime the nematic volume fraction increases until the whole sample is liquid crystalline (Figure 10.2(a)).

On passing the biphasic region the order parameter of the nematic fraction increases from a value of S=0.77 to S=0.95 through the biphasic regime, until at larger concentrations the whole sample exhibits the liquid crystalline nematic phase

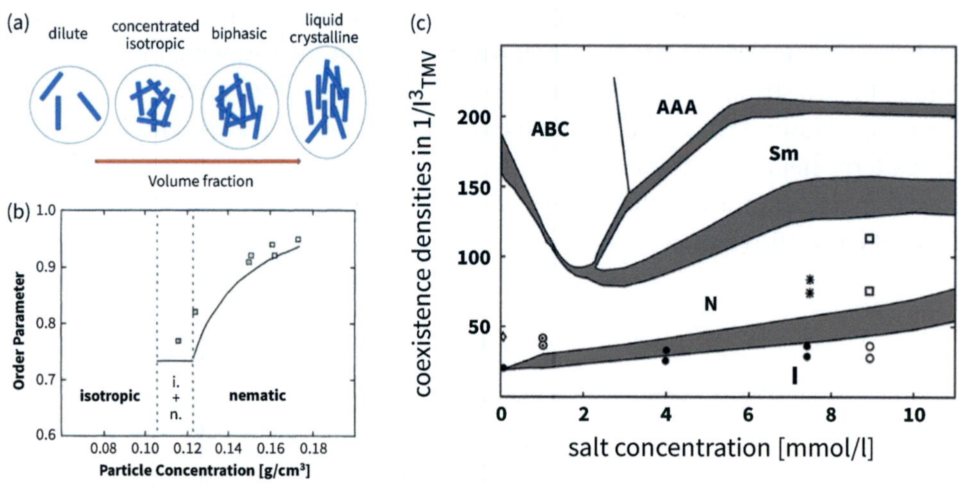

Figure 10.2 (a) Schematic illustration of the lyotropic behaviour of TMVs. By increasing concentration the isotropic phase separates into a biphasic region with increasing LC volume fraction until the whole sample exhibits the nematic phase. (b) At the same time the nematic orientational order parameter increases from S=0.77 throughout, passing the biphasic region until it reaches a value of S=0.95 far into the nematic phase [39]. (c) Complete TMV phase diagram from computer simulations in comparison to experimental values (datapoints), showing the isotropic (I), nematic (N), and smectic A (Sm) phase, together with colloidal crystal phases when varying TMV concentration and salt concentration [43].
(Figure (b) reproduced with permission from ref.[40], Figure (c) reproduced with permission from ref.[43]).

with close to perfect orientational order (Figure 10.2(b)). An explanation of why the TMV system follows Onsager's theory so well was provided by the detailed investigations of Fraden et al. [40]. They studied the birefringence of samples as a function of concentration and temperature, while also varying the polydispersity and in particular the ionic strength. This showed that the interaction between TMV rods is mainly based on electrostatic repulsion and that attractive van der Waals forces only play a rather minor role. This in turn indicates that the isotropic to nematic transition is predominantly based on excluded volume effects, one of the basic assumptions of Onsager's theory, namely the dominant repulsive steric interactions while ignoring attractive interactions.

Further X-ray experiments provided evidence of a second liquid crystalline phase, smectic A [41,42], which was also found in the computer simulations of the complete phase diagram of TMVs, together with colloidal crystal phases, presented by Graf and Löwen [43] and reproduced in Figure 10.2(c). Also, different virus suspensions have been reported to exhibit lyotropic behaviour, for example semiflexible filamentous bacteriophage with chiral nematic phases [44] and smectic ordering [45]. An overview and general outlook can be found in the more recent review articles by Dogic and Fraden [46] and Dogic [47], respectively.

10.1.2 Cellulose nanocrystals, CNC

Due to its abundance and environmental friendliness, as well as the increased interest in colloidal liquid crystals and their applications, cellulose nanocrystals (CNCs) have attracted the work of an increasing number of researchers over the last decade [48,49]. Also, cellulose nanocrystals were discussed briefly in the previous chapter on biological and living lyotropics, while here we will again concentrate on their colloidal properties playing a role in the formation of liquid crystals.

The preparation of cellulose nanocrystals progresses via the breaking down of the hierarchical arrangement of cellulose [50], as is shown in Figure 10.3. The cellulose polymer is partially crystalline, exhibiting crystal regions mediated by amorphous regions. The latter can easily be penetrated by strong acids, resulting in the hydrolysis of the polymer into CNCs. A transmission electron micrograph (TEM) [51] at the bottom right of Figure 10.3 reveals the average size of typical cellulose nanocrystals with a length of the order of 100–250 nm and a width of approximately 5–20 nm. As can be seen, the CNCs form straight, rod-like crystals with a significant polydispersity. Dimensions of CNCs depend largely on preparation methods [48,52], reaction conditions [53], and the source of the cellulose [48,49].

Due to the inherent chirality of the cellulose molecules, the CNCs form a macroscopically helical arrangement in the solvent, and the observed liquid crystalline phase is thus a chiral nematic or cholesteric phase.

Aqueous suspensions of CNCs follow the predictions of Onsager's theory, due to their lath-like shape, particularly when prepared to avoid electrostatic attraction or

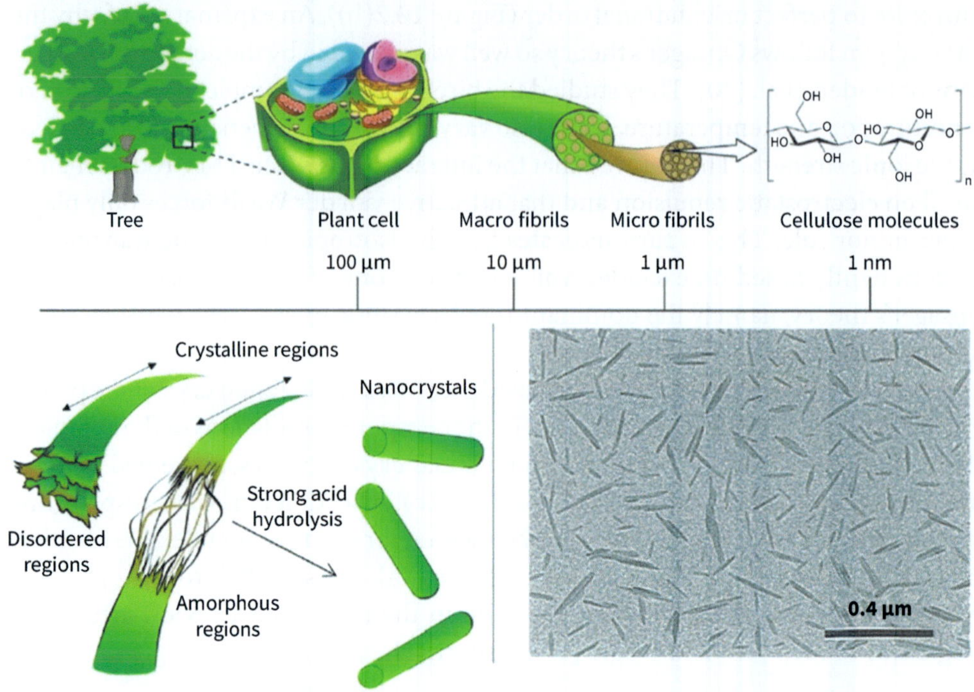

Figure 10.3 (a) Breakdown of the hierarchical structure of cellulose from natural occurrence to the molecular scale (top) [48]. The partially crystalline cellulose polymer consists of crystalline regions, mediated by amorphous, disordered regions. Chemical treatment with strong acids separated the crystalline regions to form CNCs (bottom left). (b) Typical dimensions of cellulose nanocrystals can be obtained from TEM images [51] (bottom right) to approximately 100–250 nm in length and 5–20 nm in width. CNCs are thus lath-like crystals which can form lyotropic liquid crystals.

(Figure (a) reproduced with permission from ref.[48], Figure (b) reproduced with permission from ref.[51]).

dispersion forces. If steric interactions are favoured, a similar concentration dependence as observed for other rod-like systems is found. At low concentrations of CNCs in water an isotropic, diluted phase is observed, while for increasing CNC content the volume fraction of the liquid crystalline phase increases throughout the biphasic region which is observed between approximately 3.5 and 12 wt% of CNCs. This behaviour is nicely demonstrated in Figure 10.4(a) [54] and has also been shown in [55] (Figure 10.4(b)) in sandwich preparations.

Similar to the macroscopic samples depicted in the vials of Figure 10.4(a), the passing through the biphasic region as a function of increasing CNC concentration can of course also be observed in samples between crossed polarizers in optical microscopy (Figure 10.4(b)) [55]. The textures start in the low concentration part of the biphasic region at 5 vol% with bright liquid crystalline droplets separating from the black background of the isotropic phase. Increasing the CNC concentration, the LC volume fraction increases until this is exclusively observed for a concentration of approximately 12 vol% CNC.

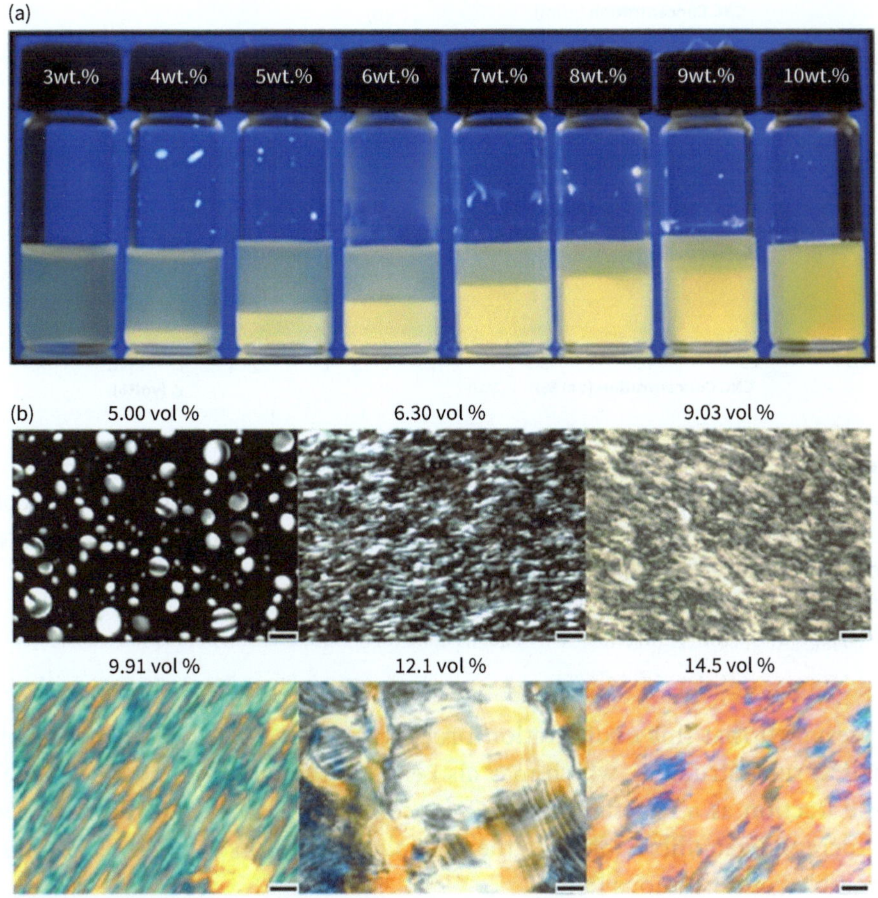

Figure 10.4 (a) Samples with increasing weight percentage of CNCs in water exhibit an increasing liquid crystal volume fraction when passing through the biphasic concentration regime [54]. (b) Similarly, this can also be observed in polarizing microscopy where the volume fraction of isotropic water decreases with increasing concentration of CNCs [55]. (Figure (b) reproduced with permission from ref.[55]).

The phase diagram of CNCs in water is slightly temperature dependent, as shown in Figure 10.5(a) which depicts the volume fraction of the liquid crystal phase as a function of CNC concentration. For low temperatures (4 °C and 20 °C) the phase diagram is practically temperature independent while for temperatures above room temperature (45 °C) the LC volume fraction is clearly lower at comparable CNC concentrations.

We mentioned above that the liquid crystalline phase of CNCs suspended in water is chiral in nature and exhibits a cholesteric helical structure. The actual phase formation is reached via equilibration. At low concentrations an isotropic suspension is observed, which for increasing CNC concentration passes through the biphasic region to form a nematic phase at higher concentrations. This nematic structure then equilibrates with time to change into the helical chiral nematic phase [57].

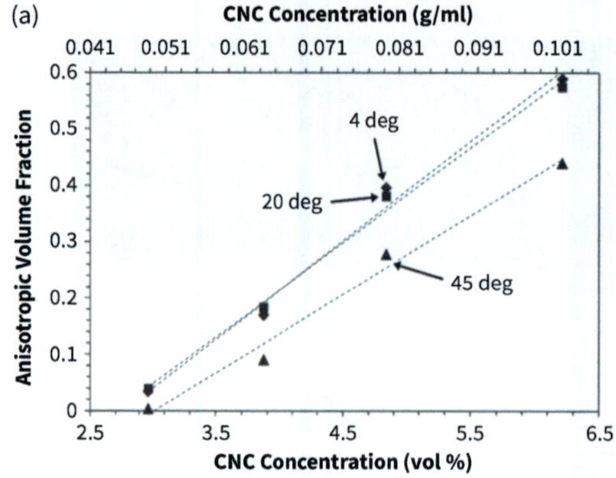

Figure 10.5 (a) At higher temperatures the CNC/water phase diagram exhibits a slight dependence on temperature with the anisotropic volume fraction decreasing at comparable CNC concentrations [55]. (b) Also the equilibration time slightly influences the phase diagram within the biphasic region, which may be due to the slow development of the helical N* superstructure from the nematic phase [56].

(Figure (a) reproduced with permission from ref.[55], Figure (b) reproduced with permission from ref.[56]).

It is presumed that this equilibration process is needed due to the long pitch of the helical superstructure, which may also explain the time dependence of the LC volume fraction as a function of CNC concentration depicted in Figure 10.5(b). With increasing equilibration time the LC volume fraction is found to decrease at comparable concentrations [56].

After the equilibration process and the formation of the helical structure of the lyotropic CNC cholesteric phase, it is found that the pitch is temperature independent [57] (Figure 10.6(a)). The dependence of the pitch on CNC concentration is on the other hand very pronounced with the pitch decreasing for increasing concentration from $P \approx 14\,\mu m$ at $\Phi \approx 3$ vol% to $P \approx 2\,\mu m$ at $\Phi \approx 6.5$ vol% (Figure 10.6(b)) [56]. This is in fact of significant importance for the application of lyotropic CNC liquid crystals because these are used to produce photonic films. To this end, the lyotropic CNC/water suspension with its helical superstructure is slowly dried, thus producing a film with increasing CNC concentration and therefore decreasing pitch and selective reflection in the visible part of the electromagnetic spectrum [58]. A dried film in a dish with diameter of about 5 cm is shown in Figure 10.6(c). At this size the film appears quite uniform except for the edges. Under microscopic enlargement (Figure 10.6(d) [59]) it can be observed that the obtained films are not quite as uniform, which affects the width of the selective reflection band [60]. The drying process and thus the uniformity of the films are effected by temperature and humidity [61], but also by shear flows [62]. The latter is of importance because the rheological properties of CNC liquid crystals are strongly dependent on concentration [63], structure [64], and ionic concentration [65].

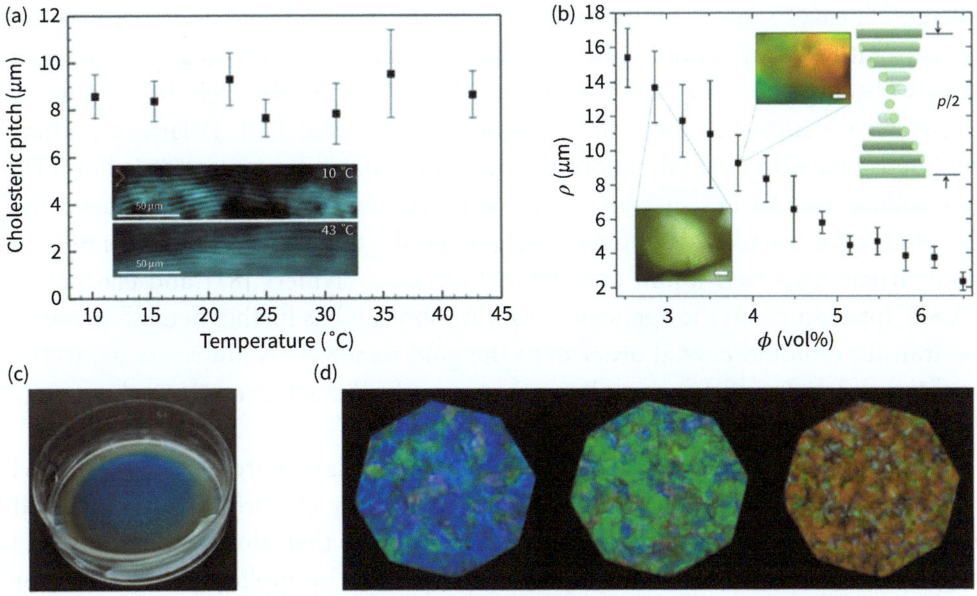

Figure 10.6 (a) The cholesteric pitch of an aqueous cellulose nanocrystal suspension is practically independent of temperature [57]. (b) On the other hand, the pitch of the helical superstructure is strongly dependent on CNC concentration [56]. (c) Drying the lyotropic liquid crystal phase results in a solid CNC film with a photonic bandgap in the visible range of the spectrum, where selective reflection is observed. (D) Microscopic view of a CNC film tuned with its reflection band from blue to green to red [59].

(Figure (a) reproduced with permission from ref.[57], Figure (c) and (d) reproduced with permission from ref.[59]).

An interesting plasmonic aspect of CNC photonic films is achieved through the addition of gold nanorods [66–68] and silver nanowires [69] to obtain co-assembled plasmonic-enhanced photonic films. Additionally, tuning can be achieved by application of electric [70] and magnetic fields [71] to the lyotropic CNC suspension, which may be maintained during film drying. For reviews of cellulose nanocrystals and more details of their synthesis, properties, and applications, we refer to the following publications [48–50, 72–77].

10.1.3 Gold nanorods

As we have seen above, the dispersion of gold nanoparticles and nanorods leads to plasmonic effects which can be very useful for applications. It is thus not surprising that the suspension of gold nanorods has so far been mainly investigated in relation to thermotropic liquid crystals. Suspending gold nanorods in a nematic phase, it has been demonstrated that the fundamental physical properties are changed significantly [78]. The conductivity σ is enhanced by orders of magnitude and also the dielectric anisotropy $\Delta\varepsilon$ is increased considerably. The threshold voltage increases slightly by the addition of gold nanorods, while the splay elastic constant K_1 nearly doubles. Besides the enhancement of physical properties, it has also been demonstrated that nanorod plasmons can effectively be modulated in an active manner through the nematic liquid crystal [79]. Polarized scattering intensities of individual nanorods can be modulated by up to 100%. For THz applications the Au-LC suspension has been shown to effectively modulate the absorbance for broad-band infrared metamaterials [80]. Other geometries of gold nanoparticle-dispersed liquid crystals include LC polymers [81] and elastomers [82,83], for example in the application to soft robots. It has further been shown that the transfer of liquid crystal order onto the gold nanorods is not restricted to the orientational order of nematic phases, but can also be achieved through smectic defects [84].

A very interesting approach to ordering gold nanorods was proposed by Umadevi et al. [85] who terminally (Figure 10.7(a)) or laterally (Figure 10.7(b)) attached mesogenic groups to the respective gold nanorods and then slowly evaporated the solvent used. These materials can then be manipulated by application of electric or magnetic fields and a Freedericksz transition was demonstrated. Self-assembly of the gold nanorods can be observed with either nematic order (Figure 10.7(c)) or smectic order (Figure 10.7(d)), which could be confirmed via small-angle X-ray scattering (SAXS). The textures observed for these materials were reported to be very similar to common nematic and smectic phases.

The orientation of Au nanorods using a lyotropic liquid crystal, in contrast to a thermotropic phase, was demonstrated by Fong et al. [86], who used a lipid-based lyotropic liquid crystal with hexagonal and lamellar phases as a matrix to suspend the nanorods. It was further shown that by near infrared (NIR) laser pulse irradiation, the nanostructure of the Au–lyotropic hybrid material could be controlled.

Liu et al. [87] reported a system of gold nanorods coated with amphiphilic materials such as CTAB (Cetyltrimethylammonium bromide) and nanoscale ionic materials (NIMs) which formed aligned self-organized systems such as illustrated in Figure 10.8(a)—a practically solvent-free lyotropic liquid crystal nanocomposite, for which the fluid-like properties result from surface functionalization and the subjection to shear. Figure 10.8(b) shows a TEM micrograph of the non-oriented gold nanorods which on application of shear display a well-oriented directionality with an order parameter 0.7 for a film containing 75% nanorods (Figure 10.8(c)). Changing the direction of shear reorients the Au nanorods into the new direction (Figure 10.8(d) and schematic Figure 10.8(e)).

A more exotic material combination of dispersing gold nanorods in a lyotropic nematic phase is the use of graphene oxide in water [88] as lyotropic (see below). The rationale of using such a system is biocompatibility, paired with electronic conduction in order to design electrode–tissue interfaces for electric stimulation and reliable control of neurons.

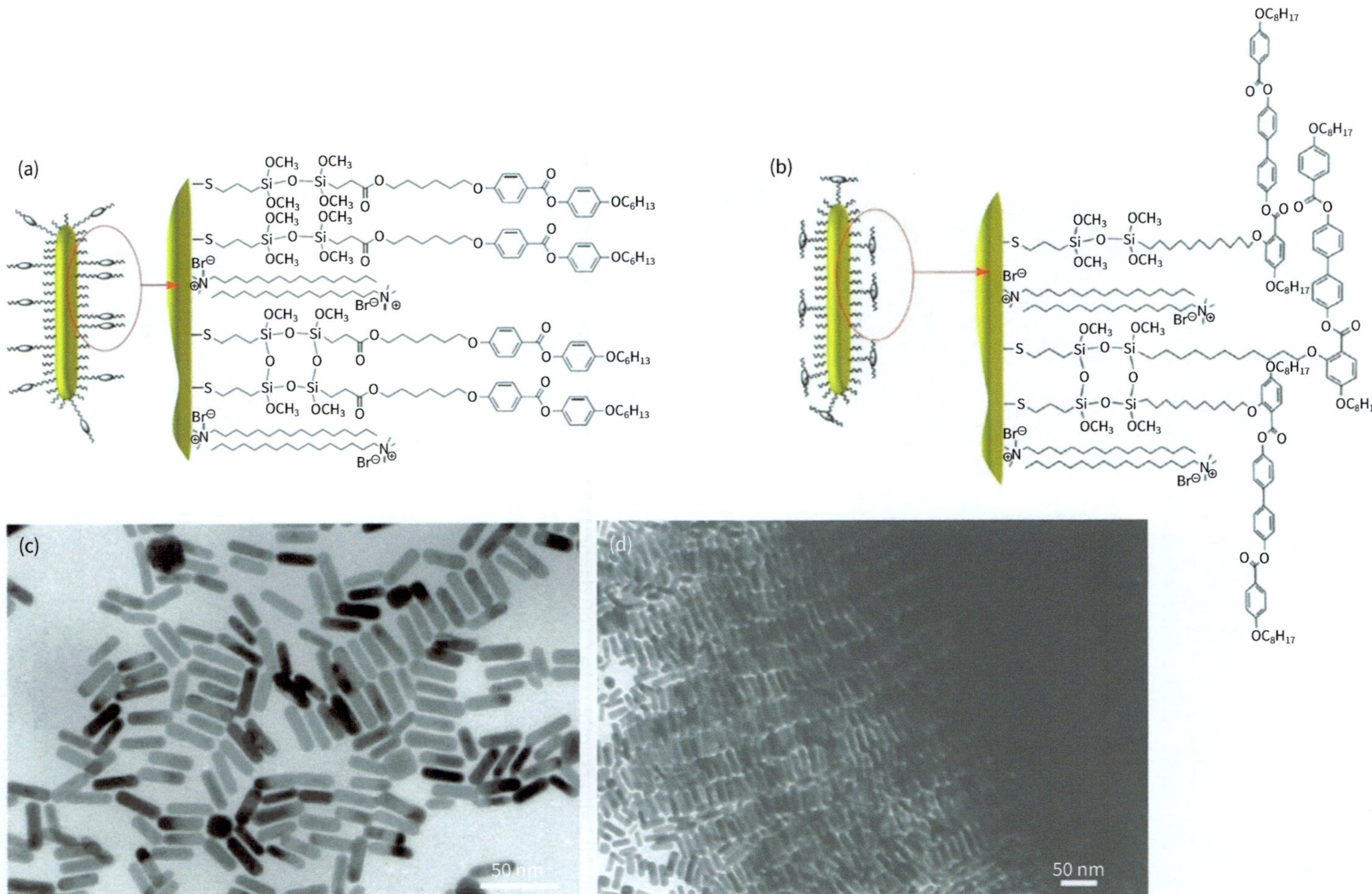

Figure 10.7 (a) Terminally attached mesogens on a CTAB decorated gold nanorod. (b) An example of laterally attached mesogen to a gold nanorod. Au-nanorods decorated in such a way can be observed to display self-organized (c), nematic, and (d) smectic order.
(Reproduced with permission from ref.[85]).

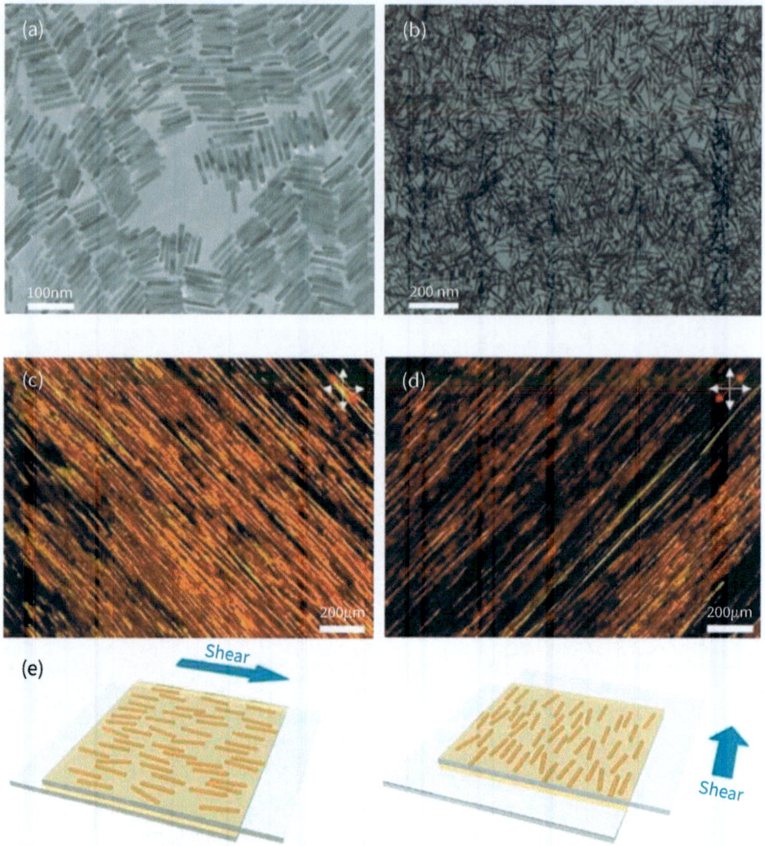

Figure 10.8 (a) Self-assembly of a CTAB decorated Au-nanorod into a smectic-like structure. (b) TEM micrograph of Au-nanorods covered with nanoscale ionic materials (NIMs) which respond to shear producing (c) well-aligned nanorod films which are stable for months. (d) Application of shear in a different direction reorients the nanorods accordingly. (e) Schematic illustration of shear switching.
(Reproduced with permission from ref.[86]).

10.1.4 Titanium dioxide, TiO_2

Besides metallic nanorods, metal oxide nanorods have also been shown to exhibit lyotropic phases, for example titanium dioxide, TiO_2. The latter is a n-type semiconductor with large bandgap of about 3.3 eV and birefringence at high refractive indices of approximately n=2.9, which makes the material attractive for opto-electronics, because it is transparent in the visible range of the spectrum. The ability to form lyotropic LCs from TiO_2 furthermore offers the possibility of manipulating and selecting optical properties by changing the orientation of the LC. The synthesis of monodisperse titanium dioxide which is capable of spontaneously forming a partially ordered lamellar structure in a solvent was reported in ref. [89]. The size of the monodisperse nanorods could be varied during the synthesis, but is generally

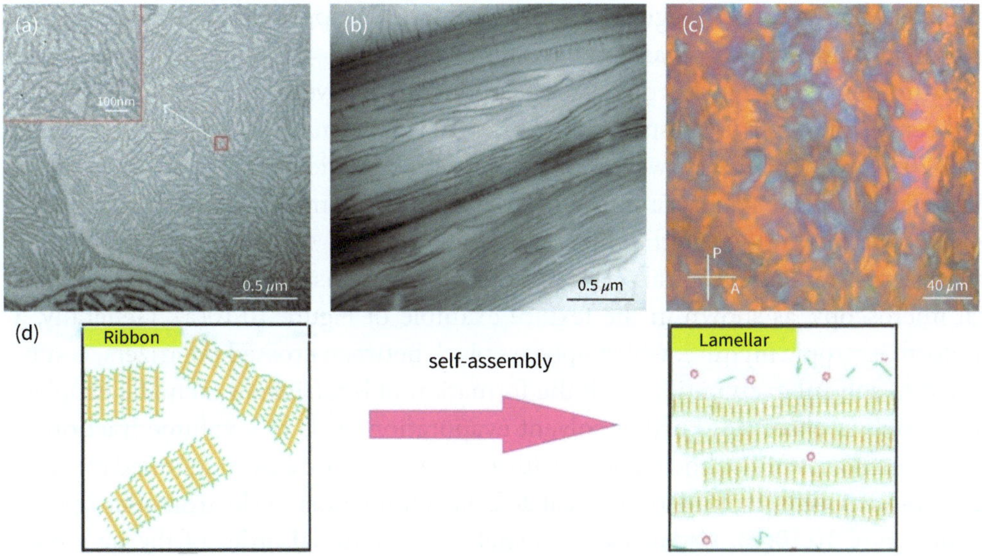

Figure 10.9 (a) TEM of titanium dioxide nanorods after the first self-assembly step, the formation of a ribbon-like structure. (b) After the second assembly step the lamellar or smectic phase is formed from the ribbons and a lyotropic LC is formed as demonstrated by (c) the texture of the lamellar phase in polarizing microscopy. (d) Schematic illustration of the two-step self-assembly process of TiO_2 nanorod lyotropic liquid crystals.
(Reproduced with permission from ref.[91]).

of the order of 30 nm in length and 4 nm in width. A further attractive route to synthesizing TiO_2 nanorods can be templating in a lyotropic liquid crystal [90]. The actual formation of a titanium dioxide nanorod lyotropic liquid crystal has reported to be accomplished via a two-step self-assembly process, which is outlined in Figure 10.9 [91].

In a first self-assembly step the titanium dioxide nanorods form a ribbon-like structure as seen in the TEM image of Figure 10.9(a). In a secondary self-assembly step these ribbons assemble in a lamellar or smectic structure as shown in Figure 10.9(b). The lamellar lyotropic liquid crystal phase exhibits clear birefringence and liquid crystalline behaviour in polarizing microscopy (Figure 10.9(c)). The formation of the lyotropic smectic order is summarized schematically in Figure 10.9(d). Effects of nanorod length, surface ligands, and inter-rod interactions, and shear and electric field alignment were discussed in detail by Hosseini et al. [92].

10.1.5 CdSe

Cadmium selenide, CdSe, is also a semiconductor, of the II–VI group. Its bandgap is only about half as wide as for TiO_2, about 1.7 eV, which falls into the red part of the visible spectrum. CdSe exhibits strong optical absorption and photoluminescence in the visible wavelength regime. Its refractive index is approximately n=2.6, which is roughly comparable to titanium dioxide.

Cadmium selenide nanorods can be produced with very good monodispersity and varied, yet controlled aspect ratio [93,94]. Typical length scales are in the range of 3–70 nm, with widths between 3 and 7 nm, which provides a range of aspect ratios between 1 and 10. The variations within a produced sample are about 5% for the width and 10% for the length [95].

The lyotropic liquid crystal phase can easily be demonstrated via polarizing optical microscopy, as shown in the texture example of Figure 10.10(a). Generally, a uniform isotropic mixture, which appears black between crossed polarizers, is subjected to solvent evaporation until the formation of birefringent, nematic droplets or tactoids is observed. Further solvent evaporation leads to a volume fraction of 100% liquid crystal, as shown above, with a nematic texture easily identified through the typical brushes around topological defects. A transmission electron micrograph, as in Figure 10.10(b), visualizes the nematic orientational order of the nanorods obtained from a Langmuir–Blodgett monolayer at slight compression [96] and is also reflected by the inset which is the Fourier transform of the TEM image. Further compression of the CdSe nanorod monolayer results in orientational and positional order as shown in the TEM image of Figure 10.10(c), resembling a smectic A phase

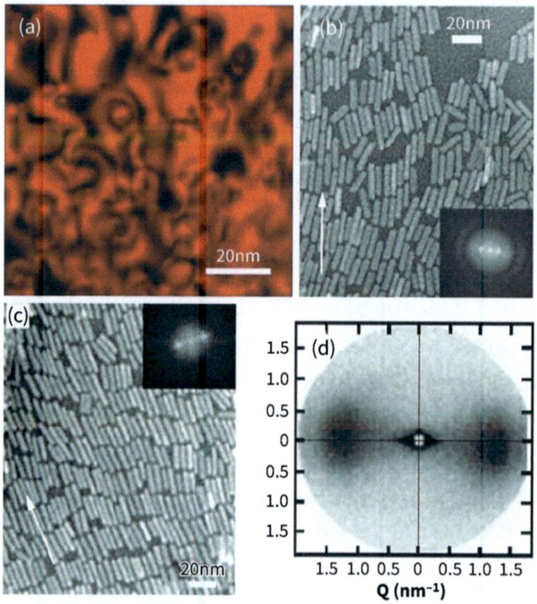

Figure 10.10 (a) Texture of the lyotropic nematic phase of CdSe nanorods [95]. (b) TEM investigations show that CdSe nanorod Langmuir–Blodgett monolayers at relatively low compression exhibit nematic orientational ordering [96]. (c) At higher compression, the nanorod ordering transforms into smectic layers [96]. (d) X-ray investigations of bulk material confirms the nematic phase of CdSe nanorods [97].

(Figure (a) reproduced with permission from ref.[95], Figure (b) and (d) reproduced with permission from ref.[96], (c) reproduced with permission from ref.[97]).

[96]. On bulk samples of capillary tubes with 300 µm diameter the spontaneous ordering of the liquid crystal phase can also be followed via wide- and small-angle X-ray diffraction. In particular, the nematic ordering is depicted in Figure 10.10(d) [97]. It was reported that the ordering of the nanorods can be optimized by adjusting the evaporation conditions and choice of the solute [98].

At this point we would also like to mention that other nanorod systems, for example silica nanorods have also been shown to exhibit lyotropic liquid crystalline behaviour. Silica nanorods have a typical length of 1 µm and a width of approximately 200 nm, while being quite polydisperse in comparison to the examples mentioned so far. Xu and Davis [99] demonstrated that silica nanorods form smectic structures in a binary solvent of dimethyl sulfoxide (DMSO) and water (H_2O). It was found that the transition from the isotropic phase to the biphasic region was strongly dependent on the concentration of DMSO, while the biphasic to LC transition was largely independent on solvent. Polydispersity had a pronounced effect on the morphology of the LC phase, showing a more uniform structure for lower polydispersity.

10.2 Nanotubes and nanowires

Just like nanorods, liquid crystal phases involving composites with nanotubes and nanowires were also mainly investigated in relation to thermotropic systems. Since over the last two decades carbon nanotubes dispersed in thermotropic nematics have been of especial interest, next we briefly summarize the general ideas behind these materials, before we discuss the work on lyotropic nanotube systems.

10.2.1 Carbon nanotubes

The investigations of thermotropic carbon nanotube dispersions originate from the idea of exploiting the spontaneous nematic self-organization to transfer orientational order onto dispersed nanotubes, either single-wall carbon nanotubes (SWCNTs) or multi-walled carbon nanotubes (MWCNTs). This was first demonstrated in 2002 by Lynch and Patrick [100], who prepared LC-ordered nanotube films and determined the order distribution function for preparations from the isotropic and the nematic phase. This distribution function was constant as a function of angle for the former case, and anisotropic for the latter, thus evidencing that the nanotubes orient along the LC director. Dierking et al. [101] showed by conductivity measurements that the nanotubes can also be dynamically reoriented when applying electric fields. Thus, they exploited the LC Freedericksz transition and elastic interactions between liquid crystals and nanotubes to switch the nanotubes from a non-conductive to a conductive state for a positive dielectric LC in planar sandwich cells (Figure 10.11(a)), or vice versa from conductive to non-conductive for a homeotropic cell and a LC with negative dielectric anisotropy (Figure 10.11(b))

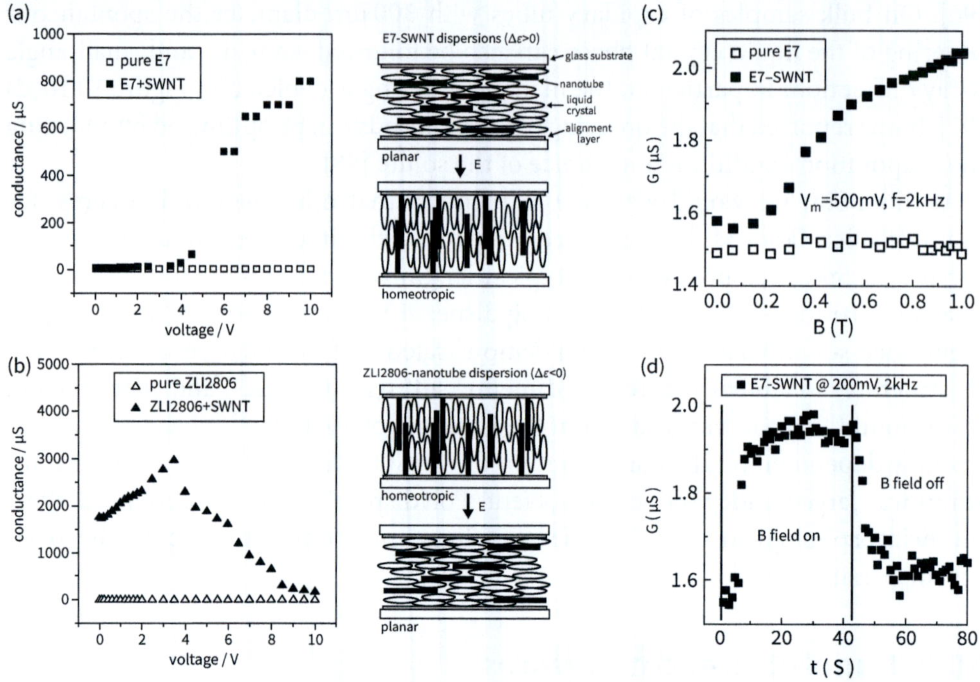

Figure 10.11 (a) The nematic liquid crystal E7, a dielectrically positive LC under planar boundary conditions, exhibits a strong increase in conductivity when dispersed with carbon nanotubes, which are conductive along their tube axis. This is due to electrically shorting the electrodes. (b) The reverse process takes place for a negative dielectric LC, ZLI2806, and homeotropic boundary conditions. Application of the electric field disrupts the conduction between the electrodes when the LC switches to a planar orientation and the carbon nanotubes follow due to elastic interactions between LC and the nanotubes. (c) A similar behaviour can be achieved through the application of magnetic fields. (d) Dynamics of the nanotube reorientation process.

(Figure (a) and (b) reproduced with permission from ref.[102], Figure (c) and (d) reproduced with permission from ref.[103]).

[102]. Similarly, such a nanotube reorientation was also demonstrated for the application of a magnetic field (Figure 10.11(c)) [103]. The dynamic experiments led to a reorientation time of the order of 20 s, which is due to the large size of the nanotubes when compared to the LC molecules (Figure 10.11(d)).

The pure liquid crystal exhibits a one-step dynamics with a single response time after removal of the electric field (Figure 10.12(a)). This switching-off time is governed by the elastic constant and the viscosity of the LC. In contrast, the LC-SWCNT dispersion exhibits a two-step relaxation with two switching-off times, one related to the LC and another related to the reorientation of the SWCNTs, elastically being driven by the director field (Figure 10.12(b)). This indicates a dynamic process which is schematically depicted in Figure 10.12(c) [104]. On the application of an electric field, the liquid crystal adopts a homeotropic director distribution and the carbon nanotubes follow this orientation, being driven through elasticity. They form

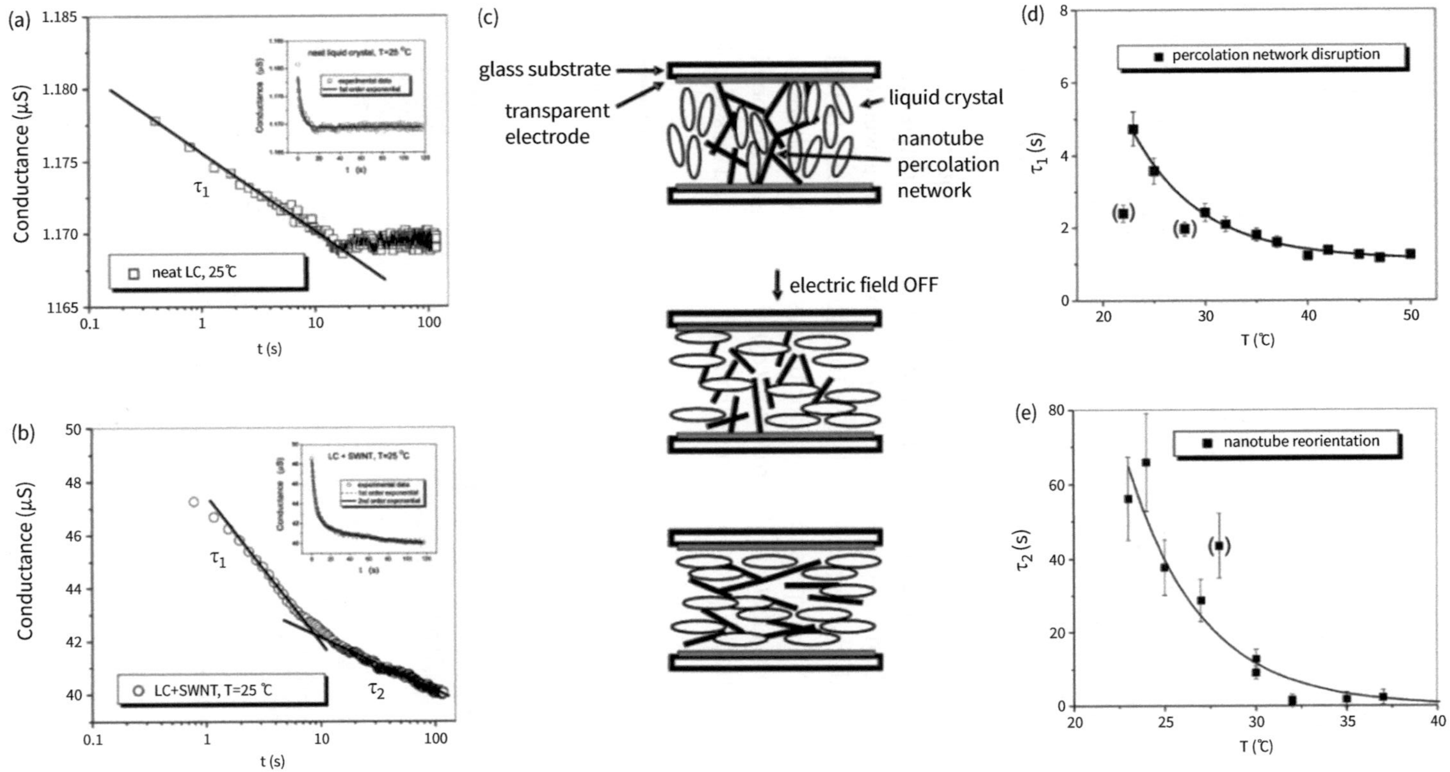

Figure 10.12 (a) Reorientation dynamics of the pure LC with a single relaxation time and (b) for the LC-SWCNT dispersion with two relaxation processes for the LC and the nanotubes. (c) Schematic illustration of the switching process by disruption of the nanotube percolation network. (d) Temperature dependence of the response time of the liquid crystal and (e) the SWCNTs of a LC-SWCNT dispersion.
(Reproduced with permission from ref.[104]).

a percolation network which leads to a conductivity much larger than that observed for the LC alone. Turning off the electric field results in a fast relaxation of the director back to the planar state with response times in the order of seconds, due to the increased viscosity of the dispersion (Figure 10.12(d)). The conductivity decreases due to the disruption of the nanotube percolation network. This is followed by a much slower relaxation of the nanotubes' reorientation to the non-conductive planar state (Figure 10.12(e)).

Besides the transfer of order of thermotropic liquid crystals onto dispersed nanotubes, this can also be achieved by dispersing CNTs in the lyotropic phase of amphiphilic molecules [105–108]. This can be realized by using different amphiphilic molecules [108] in a surfactant solution, adding carbon nanotubes and using sonification. The nanotubes will then orient in the nematic phase formed by rod-like cylinders or disc-like micelles (Figure 10.13(a)), where the tubes will then follow the nematic director. This ordering of the carbon nanotubes can be evidenced by Raman spectroscopy, the G-band, the G′-band, or the radial breathing mode (RBM) for different polarizations [106,108]. Similarly one can utilize the hexagonal lyotropic phase to macroscopically order the dispersed single-walled as well as multi-walled carbon nanotubes [108,109]. An example of a well-oriented structure is depicted with the textures of Figure 10.13(b). The relatively high viscosity of the lyotropic-CNT dispersion allows for the easy drawing of fibres, which can then be deposited on a substrate (Figure 10.13(c)). Further, a nanotube dispersion in a hexagonal lyotropic phase obtained from an ionic liquid has also been reported [110].

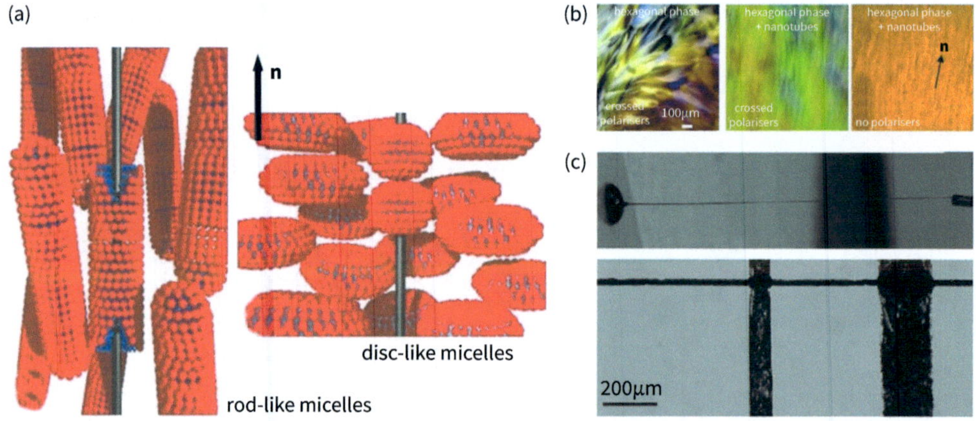

Figure 10.13 (a) Using rod-like or disc-like micelles of surfactant molecules to align the carbon nanotubes along a nematic director [105]. (b) Texture of a hexagonal lyotropic phase (left) and macroscopic alignment with dispersed nanotubes between crossed polarizers (middle) and without polarizers (right) [109]. (c) Drawing of carbon nanotube fibres and deposition on a substrate from a lyotropic LC-CNT dispersion [108].

(Figure (a) reproduced with permission from ref.[105], Figure (b) reproduced with permission from ref.[109], Figure (c) reproduced with permission from ref.[108]).

The first lyotropic phase formed from carbon nanotubes in an isotropic solvent was reported by the group of Windle [111,112]. The authors observed typical nematic Schlierentextures by polarizing microscopy with s=±1/2 topological defects (Figure 10.14(a), top row), which were also demonstrated by SEM on dried samples, exhibiting the classical director field as expected (Figure 10.14(a), bottom row). Also, for lyotropic nanotubes the phase observations follow the common Onsager description with isotropic behaviour at low concentrations, followed by a biphasic region with increasing anisotropic phase ratio for increasing nanotube concentration, until the entire sample is nematic [112].

At the same time the group of Poulin [113,114] also prepared a lyotropic nematic phase of CNTs in water by stabilizing the nanotubes via denatured DNA. Partially birefringent textures were observed in the biphasic regime at about 2–4 wt% of carbon nanotubes, while for concentrations above 4 wt% the whole sample is nematic. The phase diagram obtained via solvent evaporation is depicted in Figure 10.14(b) [113]. At the transition from the isotropic to the nematic phase, tactoids are observed [114], which are equivalent to the ones observed in many other nanorod systems, for example tobacco mosaic viruses, vanadium pentoxide, or boehmite as shown above, and are explained in [115].

It is worth noticing that it has later been demonstrated that single-walled carbon nanotubes, stabilized by double-stranded DNA, give rise to nematic as well as cholesteric phase structures with the typical oily-streaks texture and the fingerprint texture, depending on orientation of the helical superstructure [116]. The occurrence of either the nematic or the cholesteric LC structure was found to

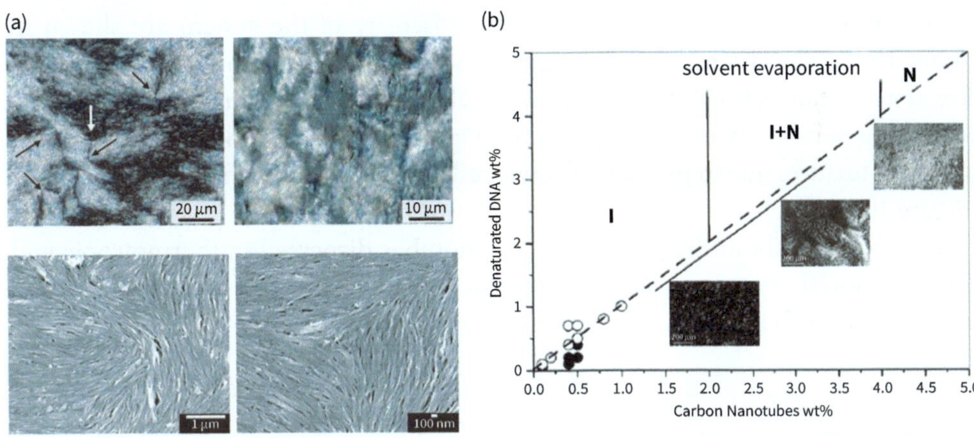

Figure 10.14 (a) Nematic textures between crossed polarizers, showing the typical Schlierentexture with brushes of the director field around the topological defects (top row). The structures of the corresponding s=±1/2 defects are depicted by SEM of dried samples (bottom row) [111]. (b) Textures and phase diagram of denatured DNS-stabilized carbon nanotubes in water, obtained by solvent evaporation [113].

(Figure (a) reproduced with permission from ref.[111], Figure (b) reproduced with permission from ref.[113]).

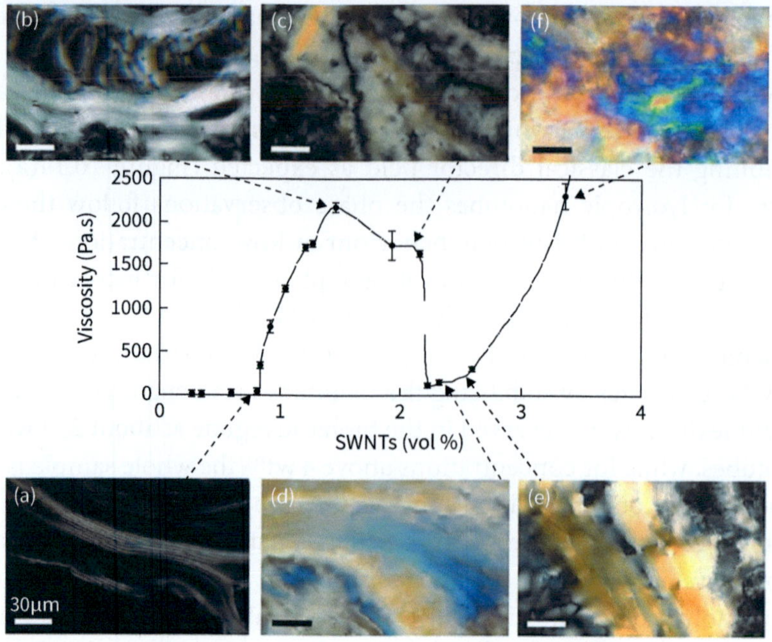

Figure 10.15 (a)—(f) Textures for increasing nanotube concentration at constant ratio of double-stranded DNA to SWCNT concentration. The pitch decreases for increasing SWCNT concentration from a resolvable fingerprint texture (a) to a texture with pitch in the visible range (f). The corresponding concentration dependence of the viscosity is shown in the central part of the figure [116].

be dependent on the relative concentration of double-stranded DNA to SWCNT. Concentration-dependent textures and the viscosity of the system are shown in Figure 10.15. The cholesteric pitch decreases with increasing nanotube concentration at a constant ratio of double-stranded DNA to SWCNT. This is illustrated by a decreasing periodicity of the fingerprints (Figure 10.15(a)) all the way to obtaining colourful textures due to pseudo-selective reflection with a pitch in the visible range (Figure 10.15(f)).

Some reviews about the properties of nanotubes dispersed in thermotropic and lyotropic liquid crystals, as well as lyotropic phases formed by carbon nanotubes, can be found in [117–120].

10.2.2 ZnO nanowires

Besides single-walled and multi-walled carbon nanotubes, there are further examples of materials forming nanotube or nanowire structures which can be used to prepare lyotropic liquid crystalline phases. One of these is zinc oxide, ZnO.

The interest in such materials is rooted in the possibility of alignment control of single-crystal nanowires for applications in semiconductor devices.

Zhang et al. [121] nicely demonstrated that ZnO nanowires, as shown in the scanning electron micrograph, SEM, of Figure 10.16(a), can possess a single-crystal structure when viewed at much larger magnification in transmission electron microscopy, TEM, like the image shown in Figure 10.16(b) of a single wire. To form liquid crystalline phases, these nanowires are decorated with a range of different ligands that promote LC behaviour (Figure 10.16(c)). The nanowires themselves are highly anisotropic with an aspect ratio of about 100 and follow Onsager's theory, as expected. At low concentration the nematic phase begins to phase separate from the isotropic liquid, passing a biphasic regime before becoming completely nematic at ZnO nanowire concentrations of approximately 5–6 vol%. This is where the typical nematic texture of Figure 10.16(d) between crossed polarizers is observed. Again, SEM investigations on dried samples show the director fields of the typical $s=\pm1/2$ topological defects (Figures 10.16(e) and (f)). In addition to mechanisms of shear alignment which can be used to orient the lyotropic phase, if the ZnO nanowires are magnetically doped with cobalt, Co, or manganese, Mn, the magnetic properties of the dopants can also be exploited to align the nanowires [122].

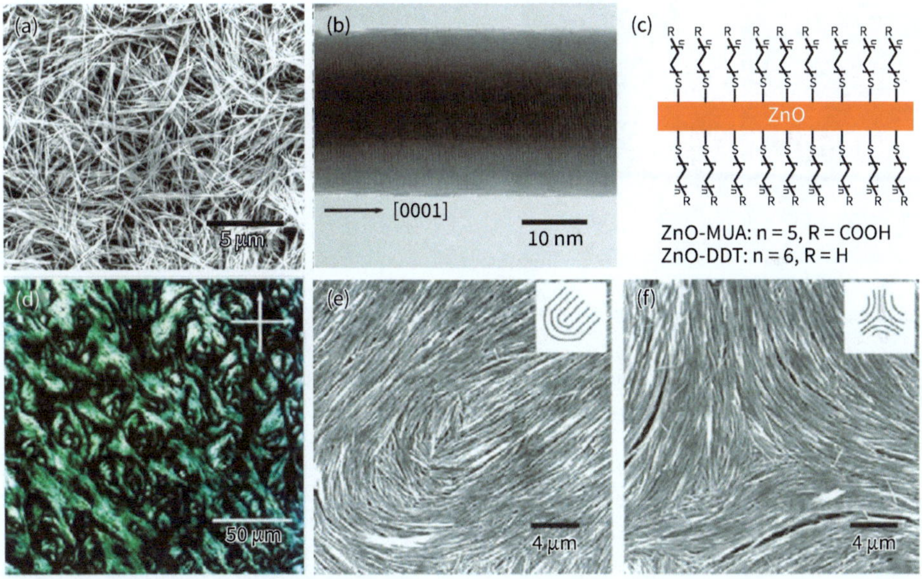

Figure 10.16 (a)SEM image of ZnO nanowires with an aspect ratio of about 100. (b) TEM image of a single-crystal ZnO nanowire. (c) Schematic illustration of various ligands on ZnO nanowires to promote liquid crystalline behaviour. (d) Nematic lyotropic phase between crossed polarizers of ZnO nanowires at a volume concentration of 6.3 vol%. (e) Dried ZnO nanowire nematic phase in the vicinity of a s=1/2 topological defect and (f) a s=−1/2 defect. (Reproduced with permission from ref.[121]).

10.2.3 Silver nanowires

In Section 10.1.3 we discussed gold nanorods, their formation of nematic and smectic phases, and their potential use in plasmonics. Similarly, one can also use other rare metals, often in the form of nanowires, for example silver, Ag. An interesting example is the mixture of silver nanowires and silver nanoparticles in an isotropic solvent, which exhibits liquid crystalline behaviour [123,124].

The system under discussion is a mixture of polyvinylpyrrolidone (PVP)-coated silver nanowires phase separated from silver nanoparticles and both dispersed in ethylene glycol or in water [124]. The Ag nanowires are highly anisotropic with an average length of approximately 4.8 μm and a diameter of about 60 nm, while the nanoparticles had a diameter of about 170 nm. As is common, the system forms an isotropic, a biphasic, and a liquid crystal regime according to Onsager's theory, which can be seen by polarizing microscopy in Figure 10.17(a) for increasing silver concentration (both nanowires and nanoparticles).

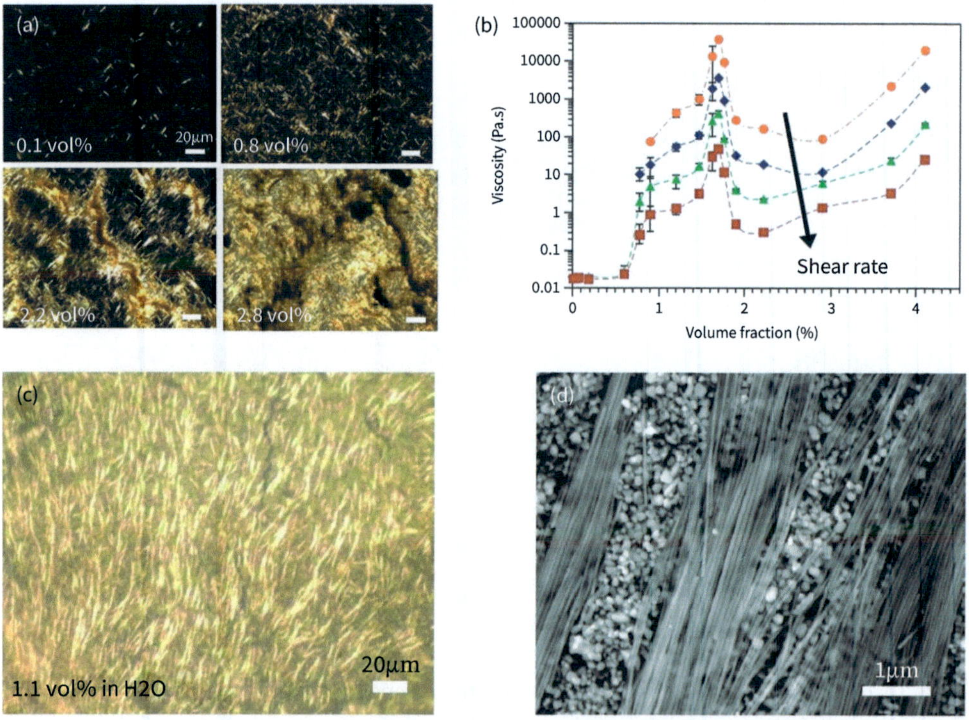

Figure 10.17 (a) Increasing volume fraction of the anisotropic phase for increasing concentration of silver content. (b) Viscosity for varying shear rate as a function of silver content through the biphasic region 0.6 vol%< c < 2.9 vol%. (c) Microscopic view of a shear-aligned lyotropic sample of Ag nanowires and nanoparticles. (d) SEM image of a dried shear-aligned sample, clearly showing the phase separation between nanowires and nanoparticles. (Reproduced with permission from ref.[124]).

For increasing shear rate the viscosity, i.e., the resistance to flow, decreases, which is a behaviour similar to that observed for LC polymers. A strong increase in viscosity is found as the system enters the biphasic region at a concentration of about 0.6 vol% silver, which is followed by a minimum once the biphasic region is passed at $c \approx 2.2$–2.9 vol%, before the viscosity once again increases as the silver concentration is further increased (Figure 10.17(b)). Shear-orientation is observed within the biphasic region at a concentration of 1.1 vol% and a shear rate of $200\,s^{-1}$, which can be observed via polarized microscopy (Figure 10.17(c)) and also in scanning electron microscopy of dried samples (Figure 10.17(d)).

Another interesting report on the orientation behaviour of silver nanoparticles dispersed in a lyotropic liquid crystal was published by Kim et al. [125]. Using atomic force microscopy (AFM), transmission electron microscopy (TEM), and electron diffraction on silver nanoparticles deposited in a lyotropic hexagonal phase, the authors demonstrated the formation of nanostrings.

10.2.4 Boron nitride nanowires, BN

A nematic lyotropic phase of boron nitride nanowires was first reported by Zhi at al. [126] for DNA-stabilized BN nanowires in aqueous solution. This was presumed by the observation of a localized nematic phase during filtration, whose volume fraction depended on the concentration of boron nitride nanowires.

Neat boron nitride nanowires, on the other hand, are not soluble in water to simply form a lyotropic phase. The solvent used by Ginestra et al. [127] was chlorosulfonic acid, which resulted in the usual isotropic-biphasic-nematic sequence for increasing BN nanowire concentration as is observed for many of the other examples discussed in more detail above. Also, the formation of tactoids was observed. The BN lyotropic liquid crystal can be extruded to form thin films, and fibres can also be drawn.

Lim et al. [128] obtained nematic phases (Figure 10.18(a)) after passing the isotropic and the biphasic region by adding different copolymeric dispersion stabilizers (Figure 10.18(b)) to the aqueous solution of BN nanowires. The phase diagrams are depicted in Figure 10.18(c) for two similar stabilizers and show the common behaviour with an isotropic phase up to a nanowire content of 9 wt%, a biphasic region in the range 9 wt% $< c <$ 14 wt% and a nematic phase above. Only the stabilizer with a short chain and an end-standing OH group shows a more pronounced liquid crystalline behaviour, lowering the isotropic to biphasic transition to 5 wt% and strongly reducing the existence range of the biphasic region, thus strongly enhancing the nematic phase. This is also shown in the POM textures of Figure 10.18(d), which show very well-developed Schlierentextures between crossed polarizers.

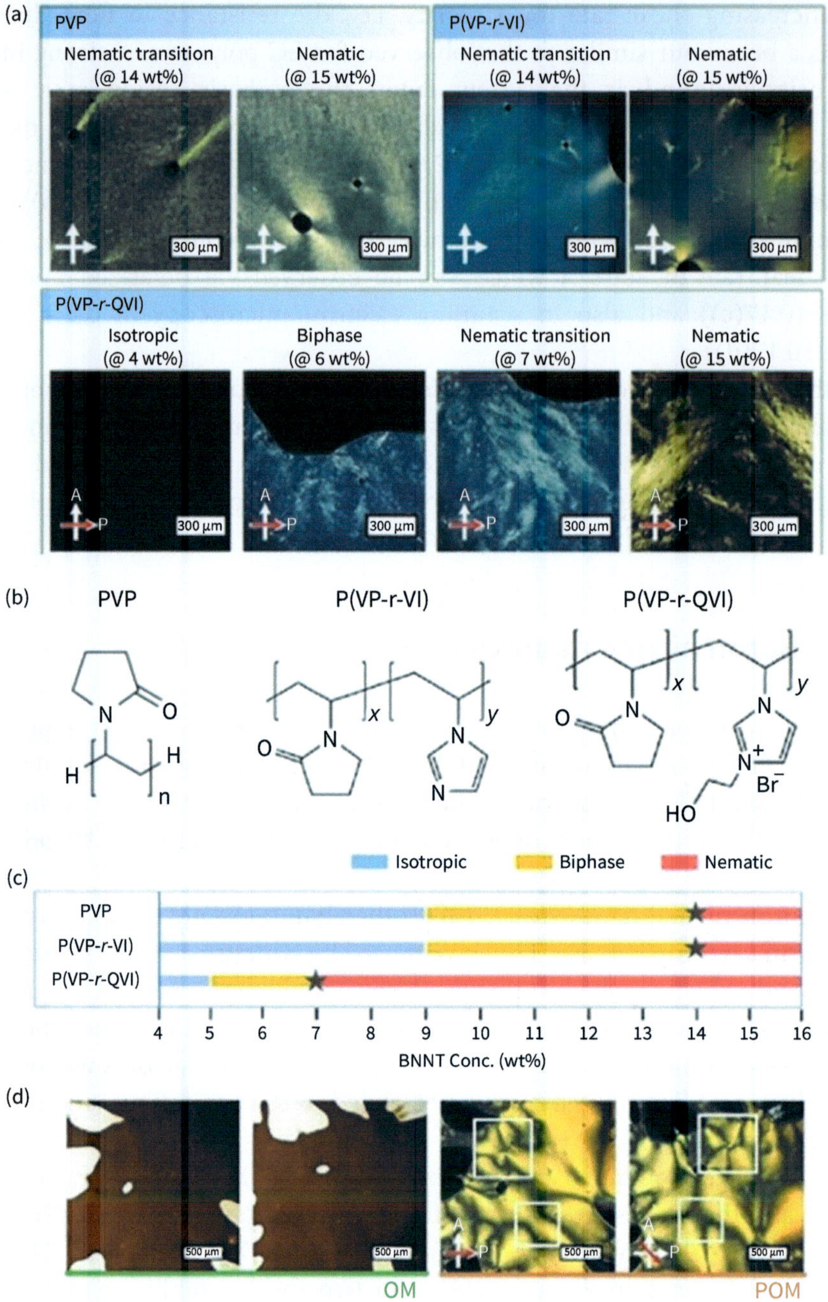

Figure 10.18 (a) Textures of stabilized aqueous solutions of boron nitride with (b) three different stabilizers. (c) Phase diagrams of the three stabilized materials, which all depict the commonly observed isotropic-biphasic-nematic behaviour. End-standing slight chains with an OH-group appear to strongly enhance the formation of the liquid crystalline phase. (d) Optical microscopy (OM) and polarized optical microscopy (POM) of the nematic textures. The POM images show a very clearly developed Schlierentexture of the BN-nanowire lyotropic liquid crystal.

(Reproduced with permission from ref.[128]).

Finally, it is worthwhile mentioning several other systems, which will not be discussed any further in relation to the formation of lyotropic liquid crystals. Besides a number of other semiconducting nanowire systems like ZnO, CdS nanoribbons in cyclohexane have also been shown to exhibit a liquid crystalline phase [129], probably with lamellar ordering.

Imogolite nanotubes in highly diluted solutions were shown to exhibit a liquid crystalline hexagonal-columnar phase [130]. This shows a very low visco-elasticity and can be aligned by application of AC electric fields.

An interesting application of lyotropic liquid crystals, although not formed by nanotubes or nanowires, is the use of the anisotropic phase as a template in the synthesis and ordering of such nanomaterials. Several nanowire materials were synthesized in rigid lyotropic surfactant phases, among them CdS [131], ZnSe [132], CuS [133], metal-boron nanotubes [124], and noble metal nanowires [135–137,123], in particular those of platinum [136,137]. A more detailed review is available by Wang et al. [138].

10.3 Nanosheets

The first disc- or sheet-like structures that were investigated with respect to liquid crystalline lyotropic behaviour were clays that we have discussed in Chapter 8. Here, we will concentrate on smaller structures and especially thinner sheet layers, often monoatomic sheets.

10.3.1 Graphene

The first monoatomic sheet structure that comes to mind is of course graphene, discovered in 2004 [13], which has been shown to exhibit interesting conductive behaviour while also being strong against tensile strains and flexible in perpendicular directions. Graphene thus represents roughly the structure of single-walled carbon nanotubes that are opened up and rolled out into a plane.

But just like carbon nanotubes, graphene does not readily dissolve in water or organic solvents, so that lyotropic behaviour is only obtained for solvents such as chlorosulfonic acid [139]. The latter reacts violently with water to form HCl gas—it is thus not the most pleasant of solvents to use. Nevertheless, at high concentrations of approximately 20–30 mg of graphene in one ml of chlorosulfonic acid, liquid crystallinity is observed (Figure 10.19) and the phase can be used to produce conductive flexible films of high transmission (above 80%), which makes it interesting for use in flexible electronics [140].

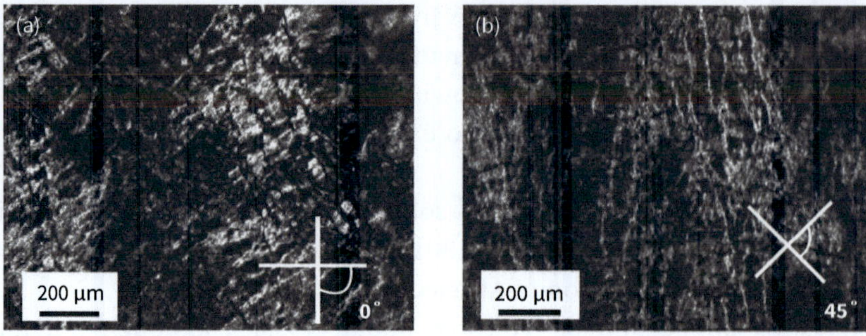

Figure 10.19 (a) Liquid crystalline phase of graphene in chlorosulfonic acid between crossed polarizers. Clear to see is the birefringence of the sample. (b) Sample as in (a) turned by 45 degrees.
(Reproduced with permission from ref.[141]).

10.3.2 Graphene oxide, GO and rGO

A sheet material very similar to graphene, namely its oxidized version, graphene oxide, GO, is much more readily transformed into a nematic lyotropic liquid crystal. GO has the same hexagonal honeycomb structure as graphene but is decorated with surface functional groups such as hydroxyl, epoxy, phenol, or carbonyl groups. These interact strongly with water and promote the formation of the liquid crystal phase.

This was realized by Kim et al. [141] and Gao et al. [142] at the same time in 2011. Like we have observed for many other lyotropic colloidal liquid crystalline systems, for GO we also have an increasing anisotropic volume fraction for increasing GO content, as shown in Figure 10.20(a) [143]. The behaviour follows the common scheme with an isotropic dispersion at low concentrations, and a biphasic region at intermediate concentrations that will finally lead into the liquid crystal phase at higher concentrations; this is schematically illustrated for GO in Figure 10.20(b). This can also be followed via POM (Figure 10.20(c)) [142] and X-ray diffraction (Figure 10.20(d)) [142].

It should be pointed out that the liquid crystal behaviour of graphene oxide can be observed in a large variety of different organic solvents (see Figure 10.21) besides water. For example Jalili et al. [145] demonstrated this for DMF (dimethylformamide), CHP (N-cyclohexyl-2-pyrrolidone), THF (tetrahydrofuran), acetone, and ethanol (Figure 10.21(a)) for all of which birefringence can clearly be observed in POM between crossed polarizers. It should also be mentioned that the phase diagram may vary and isotropic to biphasic and biphasic to LC transitions can occur at different GO concentrations for different solvents, as demonstrated by Ahmad et al. [146] in Figure 10.21(b). They also include the solvent NMP (N-methylpyrrolidone).

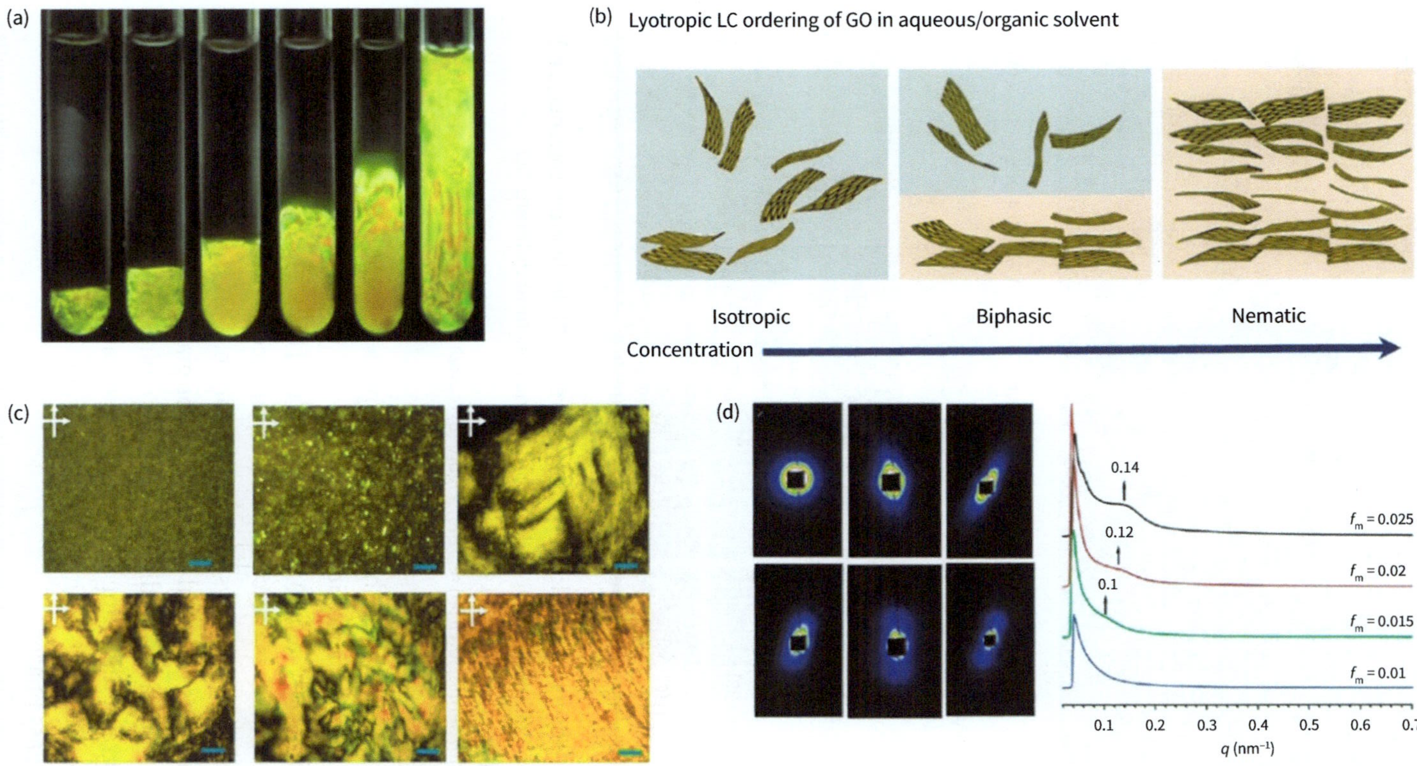

Figure 10.20 (a) Increasing anisotropic volume fraction for increasing GO concentration through the biphasic region [143]. (b) Schematic illustration of the isotropic-biphasic-nematic phase sequence for increasing concentrations of graphene oxide in an isotropic solvent [144]. (c) POM of a sample of lyotropic GO in the biphasic region (top row) and the liquid crystal state (bottom row) with a well-developed nematic Schlierentexture, for increasing concentration of GO in water [142]. (d) X-ray diffraction illustrating the transition from isotropic dispersion to nematic liquid crystal for increasing concentration of GO in water [142].

(Figure (a) reproduced with permission from ref.[143], Figure (b) reproduced with permission from ref.[144], Figure (c) and (d) reproduced with permission from ref.[142]).

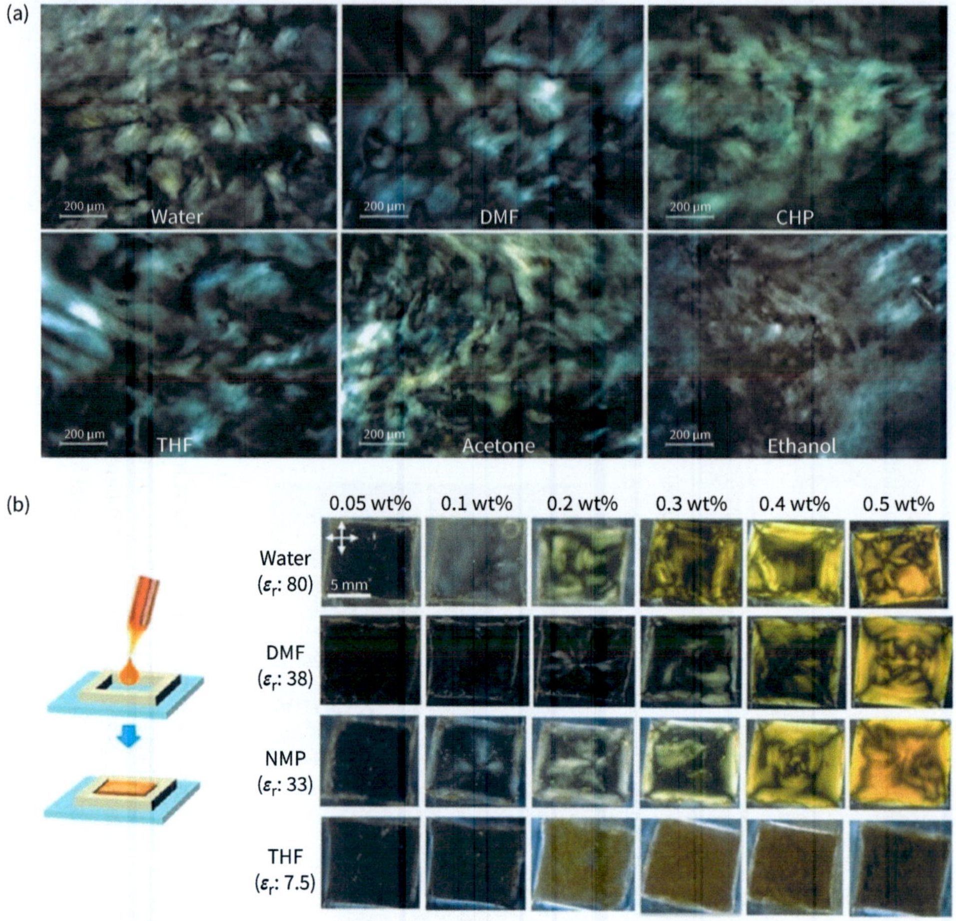

Figure 10.21 (a) POM observation of liquid crystal textures for GO in different solvents [145]. The phase diagrams may look different and transitions may occur at different GO concentrations for different solvents [146].

(Figure (a) reproduced with permission from ref.[145], Figure (b) reproduced with permission from ref.[146]).

Another factor that influences the determined phase diagram is the size of the utilized GO flakes to form a liquid crystal state, i.e., the aspect ratio of the dispersed sheets. This is clearly shown in Figure 10.22(a) [147]. The larger the GO size or aspect ratio, the smaller is the concentration needed to observe the isotropic–biphasic or the biphasic–liquid crystal phase transition. This is also the reason why a number of observations of liquid crystallinity for giant or very large GO flakes, of the size of 20–40 µm average diameter, were published [148–150]. The complete GO phase diagram, including added salt for stabilization, is shown in Figure 10.22(b) [151] and exhibits in addition to the isotropic and the nematic phase also gel, columnar, and glass phases.

The size dependence of the transition curves is also the reason why varying results were obtained for GO sources from different suppliers [141]. In the SEM diagrams of Figure 10.23(a) one can already see by eye that the average GO sheet size decreases

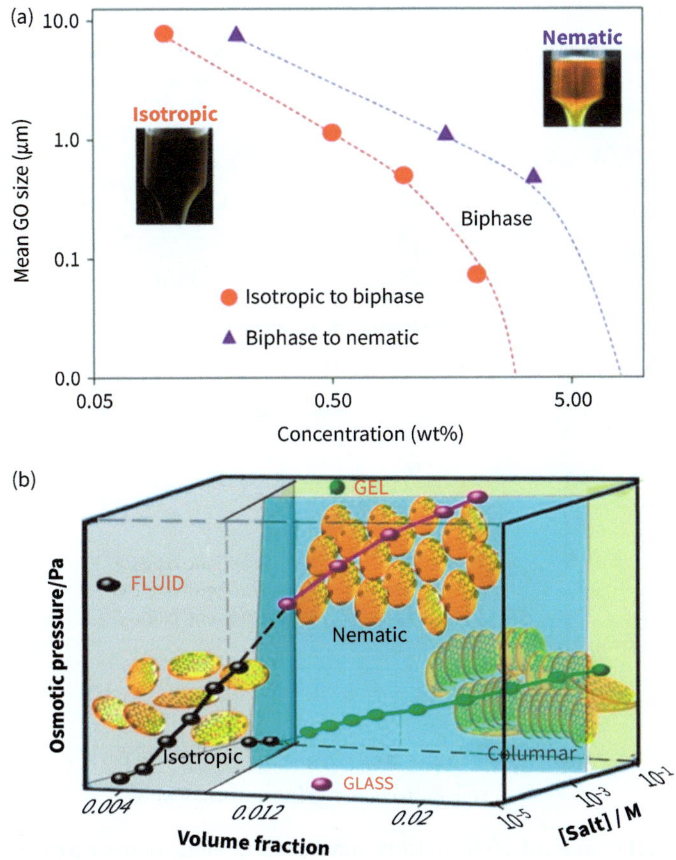

Figure 10.22 (a) Dependence of the isotropic-biphasic and biphasic-nematic transition points as a function of average size of the GO flakes. The larger the GO sheet, the lower the concentration needed to observe the respected transitions [147]. (b) Complete phase diagram of GO in an isotropic solvent [151].

(Figure (a) reproduced with permission from ref.[147], Figure (b) reproduced with permission from ref.[151]).

for samples A to C. This is reflected in the higher concentrations needed to pass the biphasic to nematic transition line, i.e., 100% volume fraction of the nematic phase at different GO concentrations (Figure 10.23(b)).

An interesting property of GO liquid crystals with respect to applications is their electro-optic switching via the Kerr effect, i.e., an induced birefringence Δn in response to an applied electric field with $\Delta n = \lambda K E^2$, where λ is the wavelength and K the Kerr constant, which is very high for lyotropic GO, of the order of 2×10^{-5} mV^{-2} (Figure 10.24(a)) [152]. A respective measurement series is shown in Figure 10.24(b) for increasing NaCl salt concentration which decreases the performance. A prototype display illustrates the potential of GO LCs in applications, although the contrast ratio is still rather small with 15:1 for the on vs the off state (Figure 10.24(c)).

Of further importance for applications is the possibility of aligning a GO liquid crystal, which can be achieved via several mechanisms. At first, a magnetic field can be applied to the LC orienting the graphene oxide sheets in the nematic state. Good

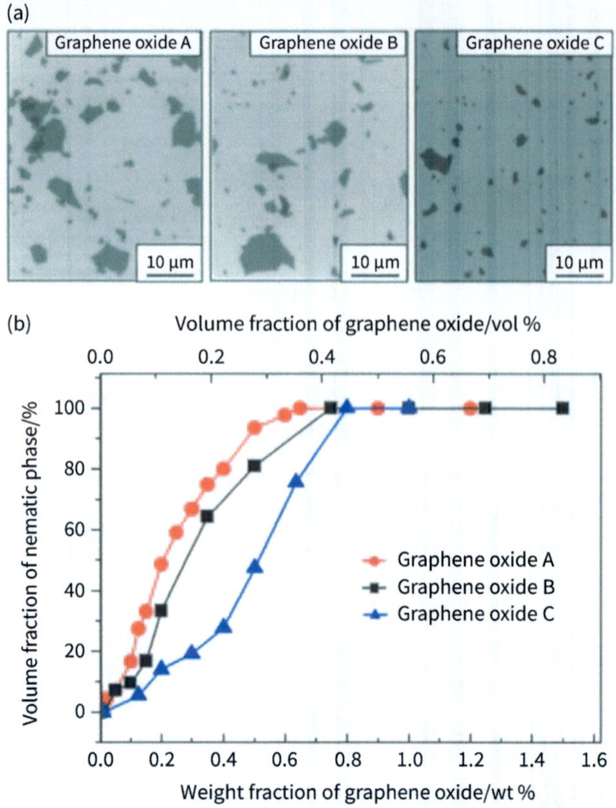

Figure 10.23 (a) SEM images of GO from three different suppliers indicate a decreasing sheet size from samples A to C. (b) For smaller sheet sizes higher volume fraction concentrations of GO are needed to obtain the same volume fraction of anisotropic phase, demonstrating that larger GO sheets are favouring the liquid crystal formation.
(Figures reproduced with permission from ref.[142]).

extinction between crossed polarizers is observed when one of the polarizers is oriented parallel to the director, while maximum transmission intensity is obtained for the polarizer at a 45° angle to the director (Figure 10.25(a)). Good alignment can be obtained over large sample areas [141].

Further, bounding glass plates, for example of a microchannel as shown in Figure 10.25(b), can be used to surface align the GO-LC without any treatment of the substrates. The GO sheets prefer to lie parallel to the bounding substrates, which leads to the bright appearance in the open microchannel in the right-hand side of Figure 10.25(b) with only a slight splay deformation leading to a small variation in transmitted intensity. On the left-hand side, the channel is closed with a top substrate, which leads to a substantial splay deformation for GO sheets coming from both sides of the channel. Consequently, the director adopts a homeotropic orientation in the middle of the channel while pointing at a 90° angle on either side, leading to a dark stripe in the middle and high transmission on the sides. This demonstrates clearly the surface alignment possibilities of GO liquid crystals [153].

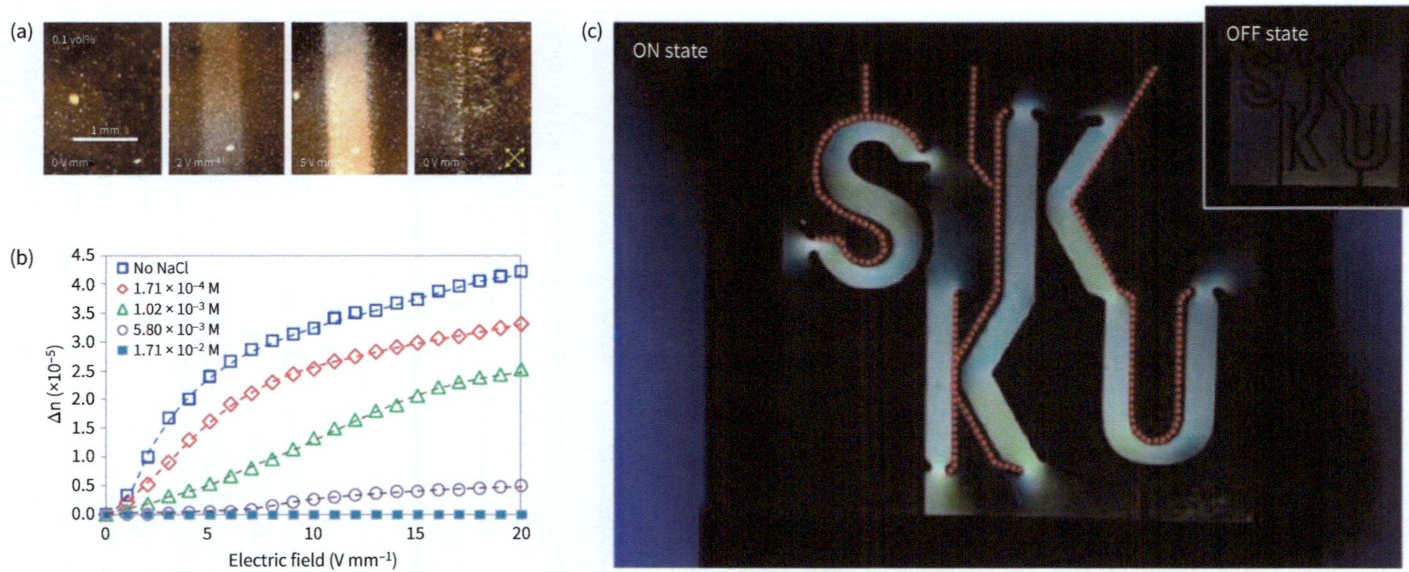

Figure 10.24 (a) Qualitative illustration of the Kerr effect switching of a GO-LC in the dark state at 0 V mm^{-1}, increasing birefringence at 2 and 5 V mm^{-1}, before switching back to the off state. (b) The addition of NaCl salt decreases the Kerr performance of the GO-LC. (c) GO-LC Kerr effect-based display with an aspect ratio of 15:1 between the on and the off state.

(Reproduced with permission from ref.[152]).

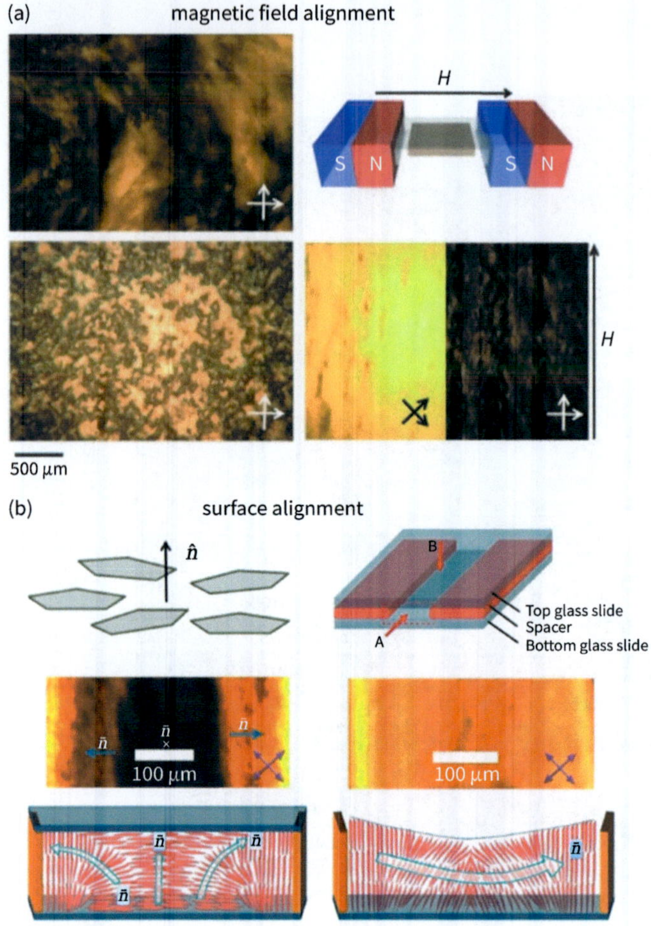

Figure 10.25 (a) Magnetic field alignment of a GO nematic liquid crystal demonstrated by a well-aligned director field with good extinction between crossed polarizers [142]. (b) Surface alignment of a GO nematic LC in a microchannel under different boundary conditions, yet with untreated glass substrates [153].

(Figure (a) reproduced with permission from ref.[142], Figure (b) reproduced with permission from ref.[153]).

Similar to the drawing of carbon nanotube fibres from a lyotropic LC-CNT dispersion as discussed above, fibres of graphene oxide can also be drawn from the liquid crystalline state. This is demonstrated in Figure 10.26 [154]. The original fibre can be viewed between crossed polarizers and one can see that some of the birefringence is retained from the liquid crystallinity confined in the voids (Figure 10.26(a)). Part (b) of the figure, an SEM image, depicts the almost circular diameter of the fibre, in which the graphene oxide sheets are folded as seen in the cross-section enlargement of Figure 10.26(c). The fibres are flexible enough to be knotted and can be annealed to produce reduced graphene oxide (rGO), to regain some of the conductive properties that may be beneficial for applications of nanowires (Figure 10.26(d)).

GO fibres can be drawn that are several metres in length (Figure 10.27(a)). These can be woven into a fabric as shown in Figure 10.27(b) as black fibres from left to

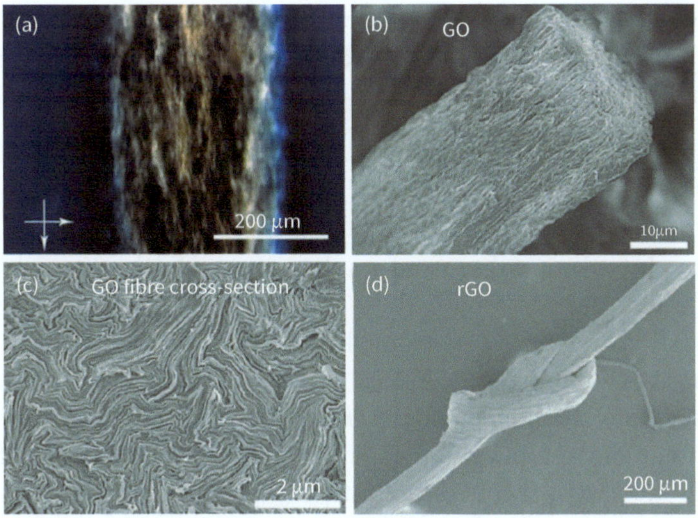

Figure 10.26 (a) A drawn fibre between crossed polarizers showing residual birefringence through trapping some of the liquid crystal in micrometre-sized voids. (b) An SEM image of the GO fibre exhibits a near circular diameter. (c) Fibre cross-section with folded GO sheets. (d) The drawn fibres are flexible enough to be knotted and can also be annealed to form reduced graphene oxide, rGO.
(Reproduced with permission from ref.[154]).

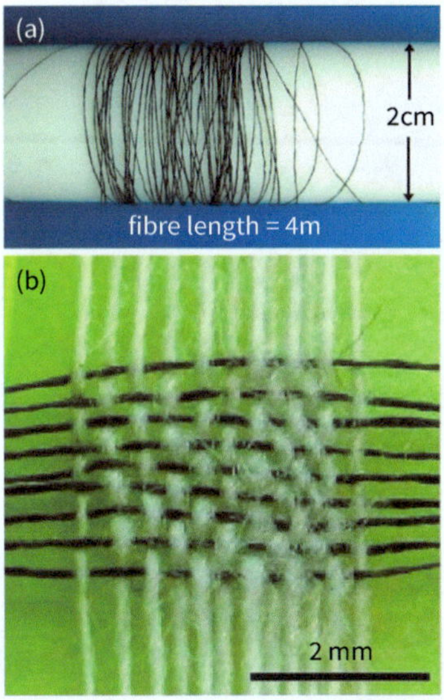

Figure 10.27 (a) A spindle with a continuously drawn GO fibre from the LC state of length of 4 m. (b) The fibre can be woven into a fabric or combined with other fibres like cotton.
(Reproduced with permission from ref.[155]).

right, together with white cotton fibres running from top to bottom [155] and can also be reduced to rGO. Other fibre-drawing demonstrations have been reported [156,157] and were reviewed in [158,159].

Since graphene oxide liquid crystals have by now been a field relatively well researched, there are several reviews available. For more detailed descriptions and lists of reference we refer the reader to the review papers of [143,144,160,161].

10.3.3 MoS_2

A similar example to the just discussed graphene oxide are the dispersions of MoS_2 nanosheets in water, as reported by Jalili et al. [162]. At a certain concentration of approximately 0.1 mg ml^{-1} a lyotropic liquid crystal phase begins to be formed and the biphasic regime is reached. At a concentration of about 0.3 mg ml^{-1} the nematic phase is fully developed and observed in POM, which becomes even more pronounced after shear application (Figure 10.28(a)). The viscosity of the phase increases with increasing sheet concentration (Figure 10.28(b)). A small decrease is found at concentrations where the nematic phase is formed (0.3 mg ml^{-1}), which is due to the slight increase in free volume provided by the ordering of the MoS_2 nanosheets.

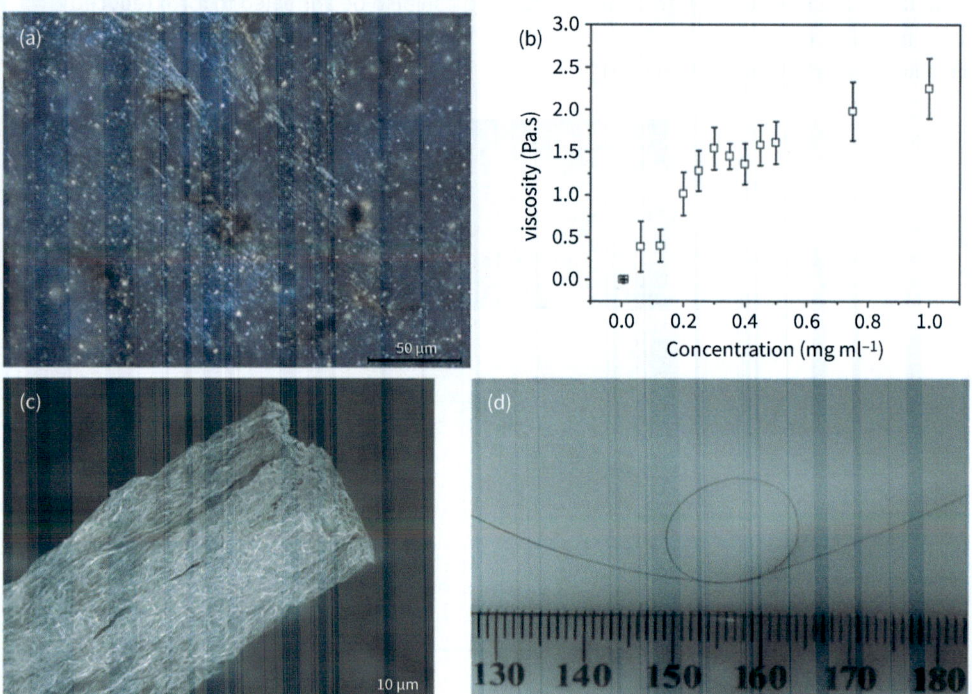

Figure 10.28 (a) Texture of the anisotropic MoS_2 liquid crystal phase after shear at a concentration of approximately 0.3 mg ml^{-1}. (b) Concentration dependence of the viscosity of the MoS_2 nematic through the isotropic phase, the biphasic regime (0.1–0.3 mg ml^{-1}) and above in the nematic phase. (c) SEM of a drawn fibre from a MoS_2 liquid crystal, which can (d) be obtained at considerable length.
(Reproduced with permission from ref.[162]).

The MoS_2-LC is also viscous enough for the drawing of fibres, as can be seen in the scanning electron micrograph of Figure 10.28(c). These can be drawn continuously to extend over considerable length (Figure 10.28(d)).

10.3.4 MXenes

MXenes are two-dimensional materials which consist of a thin atomic layer of transition metal nitrides, carbides, or carbonitrides and were first described in 2011 [163]. They have been synthesized in three distinct groups: (i) M_2XT_x, (ii) $M_3X_2T_x$, and $M_4X_3T_x$. In these groups M stands for the transition metal (for example Ti, V, Cr, Zr, Nb, Mo), X for either C and/or N, and T for a possible surface termination, which is O, F, or Cl. This allows for a large variation of nanosheet materials which are discussed in more detail in [164]. Main applications are envisioned in the fields of sensors, energy storage, catalysts, or electromagnetic shielding.

MXenes in water undergo the same sequence of transitions for increasing nanosheet concentrations as all the other Onsager-based systems, from an isotropic dispersion through a biphasic region into a fully anisotropic liquid crystal phase, which was reported for the first time in 2020 by Zhang et al. [165,166]. An illustrative POM texture series is depicted in Figure 10.29(a) for the MXene Ti_3C_2 at different indicated concentrations [167]. MXene sheet sizes were relatively small of

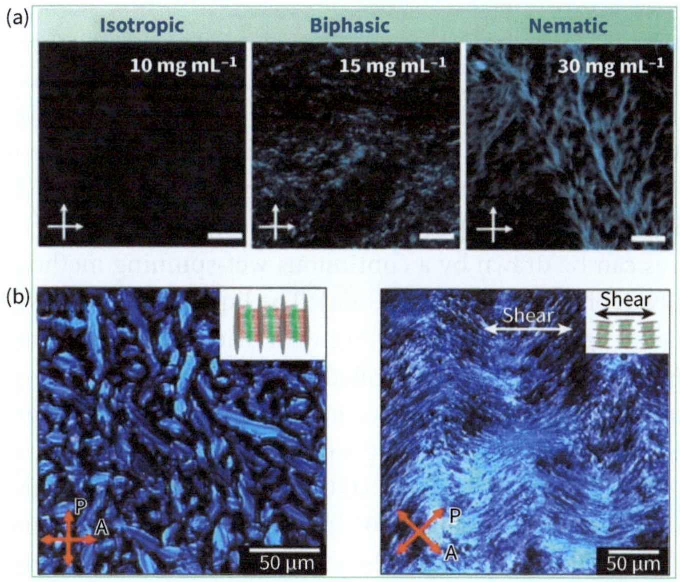

Figure 10.29 (a) The typical sequence of isotropic dispersion—biphasic region—anisotropic phase as observed for all Onsager systems [167]. (b) The developing anisotropic phase exhibits a fan-like texture which indicates the formation of a lamellar phase, which can be reoriented shear application [168].

(Figure (a) reproduced with permission from ref.[167], Figure (b) and (c) reproduced with permission from ref.[168]).

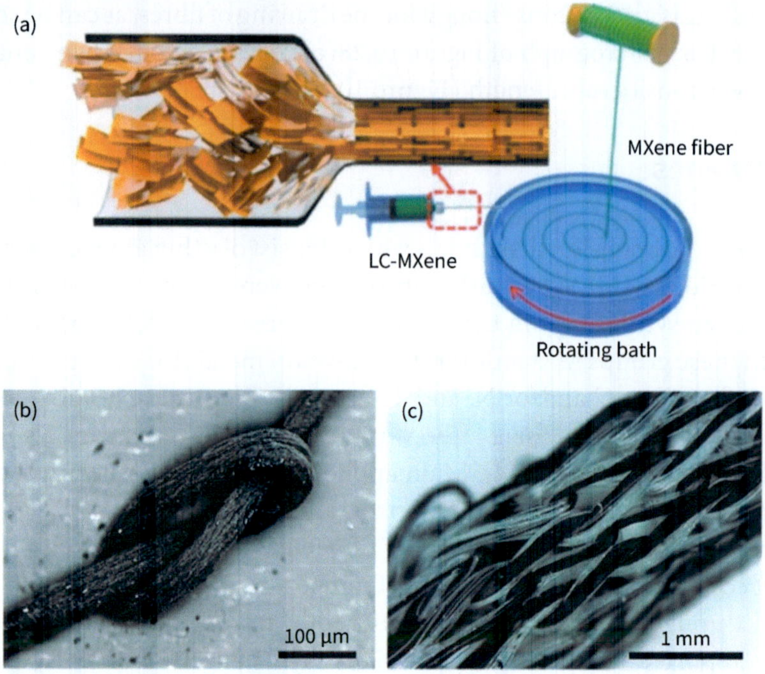

Figure 10.30 (a) Schematic illustration of the wet-spinning process employed to produce MXene fibres[169]. (b) A fibre produced from 88% MXene/12% GO, as an example of a mixed fibre to add functionality and optimize properties. (c) Co-knitted MXene-based fibre, MXene-GO, with Nylon yarn [170].

(Figure (a) reproduced with permission from ref.[169], Figure (b) and (c) reproduced with permission from ref.[170]).

dimensions, in the µm regime. Additionally, the paper also discusses mixtures of MXene and CNCs. The POM texture shows a developing fan-like texture which indicates the formation of a smectic/lamellar structure (Figure 10.29(b)), which can be reoriented through shear application [168].

MXene fibres can be drawn by a continuous wet-spinning method [169], which is schematically depicted in Figure 10.30(a). The liquid crystalline MXene phase is pressed through a syringe, forming a fibre with oriented MXene sheets. This fibre is deposited into a rotating bath and rolled up continuously onto a spool. The dispersed MXene can be mixed with other nanosheet structures like graphene oxide to add functionality and to optimize properties. This produces mixed fibres like in Figure 10.30(b) for 88%MXene-12%GO which demonstrates the possibility of knotting due to the fibre flexibility [170]. These fibres can be combined with other fibres through co-knitting, for example MXene/GO fibres co-knitted with Nylon yarn (Figure 10.30(c)).

References

[1] Y. Shen and I. Dierking (2019). Perspectives in liquid-crystal-aided nanotechnology and nanoscience. *Appl. Sci.*, 9, 2512. doi:10.3390/app9122512

[2] R. Manohar, S. P. Yadav, A. K. Srivastava, A. K. Misra, K. K. Pandey, P. K. Sharma, and A. C. Pandey (2009). Zinc oxide (1% Cu) nanoparticle in nematic liquid crystal: dielectric and electro-optical study. *Jpn. J. Appl. Phys.*, 48, 101501.

[3] S. Mishra, V. Manjuladevi, and R. K. Gupta (2023). Effect of shape of ZnO nanoparticle on electro-optic and dielectric properties of nematic liquid crystal. *J. Mol. Liq.*, 386, 122482.

[4] W.-T. Chen, P.-S. Chen, and C.-Y. Chao (2009). Effect of doped insulating nanoparticles on the electro-optical characteristics of nematic liquid crystals. *Japn. J. Appl. Phys.*, 48, 015006.

[5] A. Kanwar, S. Ghodke, and V. Gajghate (2023). Determination of refractive index and birefringence of nanoparticle doped liquid crystals. *Internat. J. Thermophys.*, 44, 35. https://doi.org/10.1007/s10765-022-03145-3

[6] Khushboo, A. Devi, P. Malik, H. Kumar (2016). Thermodynamic and acoustical study of zinc oxide-nematic liquid crystals mixtures. *J. Mol. Liq.*, 214, 145–148.

[7] P. K. Tripathi, A. K. Misra, K. K. Pandey, S. P. Yadav, R. Manohar (2013). Abnormal switching behavior of nanoparticle composite systems. *Phase Transitions*, 86, 1241–1255. doi: 10.1080/01411594.2012.761698

[8] A. Chaudhary, P. Malik, R. Mehra, and K. K. Raina (2013). Influence of ZnO nanoparticle concentration on electro-optic and dielectric properties of ferroelectric liquid crystal mixture. *J. Mol. Liq.*, 188, 230–236.

[9] D. P. Singh, S. K. Gupta, T. Vimal, and R. Manohar (2014). Formation of periodic domains and change in physical properties of paramagnetic copper doped ZnO nanoparticle dispersed ferroelectric liquid crystal system. *J. Mol. Liq.*, 198, 267–273.

[10] G. Pathaka, R. Katiyara, K. Agraharia, A. Srivastavaa, R. Dabrowskib, K. Garbatb, and R. Manohar (2018). Analysis of birefringence property of three different nematic liquid crystals dispersed with TiO_2 nanoparticles. *Opto-Electronics Review*, 26, 11–18.

[11] C. Y. Huang, P. Selvaraj, G. Senguttuvan, and C. J. Hsu (2019). Electro-optical and dielectric properties of TiO_2 nanoparticles in nematic liquid crystals with high dielectric anisotropy. *J. Mol. Liq.*, 286, 110902.

[12] H. Ayeb, S. Alaya, M. Derbali, L. Samet, J. Bennaceur, F. Jomni, and T. Soltani (2021). Dielectrical, electro-optical and textural studies of 5CB nematic liquid crystal doped with TiO_2 and Cu-TiO_2 nanoparticle. *Liq. Cryst.*, 48, 223–232. doi: 10.1080/02678292.2020.1771784

[13] G. Pathaka, S. Pandeya, R. Katiyara, A. Srivastavaa, R. Dabrowskib, K. Garbatb, and R. Manohar (2017). Analysis of photoluminescence, UV absorbance, optical band gap and threshold voltage of TiO_2 nanoparticles dispersed in high birefringence nematic liquid crystal towards its application in display and photovoltaic devices. *Journal of Luminescence*, 192, 33–39.

[14] M. Sharma, A. Sinha, and M. R. Shenoy (2015). Effect of TiO_2 nanoparticle doping on the performance of electrically controlled nematic liquid crystal core waveguide switch. *Optical Materials*, 49, 292–296.

[15] C. J. Hsu, B. P. Singh, P. Selvaraj, M. Antony, R. Manohar, and C. Y. Huang (2021). Superior improvement in dynamic response of liquid crystal lens using organic

and inorganic nanocomposite. *Scientific Reports*, 11, 17349. https://doi.org/10.1038/s41598-021-96991-4

[16] X. Wang, W. Hu, G. Chen, H. Chen, R. Huang, Y. Ren, J. Xu, J. Hu, L. Zhang, M. Yu, H. Zhang, and H. Yang (2022). TiO_2 doped polymer dispersed and stabilised liquid crystal smart film with high contrast ratio, low driving voltage and short response time. *Liq. Cryst.*, 49, 1623–1632. doi: 10.1080/02678292.2022.2048912

[17] W.-K. Lee, J.-H. Choi, H.-J. Na, J.-H. Lim, J.-M. Han, J.-Y. Hwang, and D.-S. Seo (2009). Low-power operation of vertically aligned liquid-crystal system via anatase-TiO_2 nanoparticle dispersion. *Opt. Lett.*, 34, 3653–3655.

[18] R. S. Pathinti, B. Gollapelli, S. K. Jakka, and J. Vallamkondu (2021). Green synthesized TiO_2 nanoparticles dispersed cholesteric liquid crystal systems for enhanced optical and dielectric properties. *J. Mol. Liq.*, 336, 116877.

[19] P. Kumar and A. Sinha (2017). Effect of TiO_2 nanoparticle doping on the electrical properties of ferroelectric liquid crystal. In: I. Bhattacharya et al. (eds), *Advances in Optical Science and Engineering*, Springer Proceedings in Physics, 194, 499–505. doi: 10.1007/978-981-10-3908-9_61

[20] N. Yadav, R. Dabrowski, and R. Dhar (2014). Effect of alumina nanoparticles on dielectric permittivity, electrical conductivity, director relaxation frequency, threshold and switching voltages of a nematic liquid crystalline material. *Liq. Cryst.*, 41, 1803–1810. doi: 10.1080/02678292.2014.950619

[21] H. Karabuğa, H. H. Mert, G. Karanlık, and H. Ocak (2019). Synthesis and characterization of bent-core liquid crystal/modified γ-Al_2O_3 nanocomposites. *ChemistrySelect*, 4, 8983–8988. doi: 10.1002/slct.201902589

[22] S. Kumar, S. Khatua, and P. Thareja (2019). Fumed alumina-in-nematic liquid crystal suspensions under shear and electric field. *Rheologica Acta*, 58, 203–216. https://doi.org/10.1007/s00397-019-01132-4

[23] M. Gupta, I. Satpathy, A. Roy, and R. Pratibha (2010). Nanoparticle induced director distortion and disorder in liquid crystal-nanoparticle dispersions, *J. Colloid Interface Sci.*, 352, 292–298.

[24] M. Hasegawa (2012). Characterization of $BaTiO_3$ nanoparticles suspension in liquid crystal. *J. Photopoly. Sci. Technol.*, 25, 296–299.

[25] G. Cook, V. Yu. Reshetnyak, R. F. Ziolo, S. A. Basun, P. P. Banerjee, and D. R. Evans (2010). Asymmetric Freedericksz transitions from symmetric liquid crystal cells doped with harvested ferroelectric nanoparticles. *Optics Express*, 18, 17339.

[26] F. Li, J. West, A. Glushchenko, C. I. Cheon, and Yu. Reznikov (2006). Ferroelectric nanoparticle/liquid-crystal colloids for display applications, *J. SID*, 14/6, 523–527.

[27] K. P. Sigdel and G. S. Iannacchione (2013). Calorimetric study of phase transitions in ocylcyanobiphenyl-barium titanate nanoparticle dispersions. *J. Chem. Phys.*, 139, 204906.

[28] R. Basu (2014). Soft memory in a ferroelectric nanoparticle-doped liquid crystal. *Phys. Rev. E*, 89, 022508.

[29] G. Cook, A. V. Glushchenko, V. Reshetnyak, A. T. Griffith, M. A. Saleh, and D. R. Evans (2008). Nanoparticle doped organic-inorganic hybrid photorefractives. *Optics Express*, 16, 4015–4022.

[30] E. Mavrona, U. Chodorow, M. E. Barnes, J. Parka, N. Palka, S. Saitzek, J.-F. Blach, V. Apostolopoulos, and M. Kaczmarek (2015). Refractive indices and birefringence of hybrid liquid crystal - nanoparticles composite materials in the terahertz region. *AIP Advances*, 5, 077143.

[31] R. Uttam, S. Kumar, and R. Dhar (2020). Magnified charge carrier conduction, permittivity, and mesomorphic properties of columnar structure of a room temperature discotic liquid crystalline material due to the dispersion of low concentration ferroelectric nanoparticles. *Phys. Rev. E*, 102, 052702.

[32] A. Kumari and A. Sinha (2021). Role of $BaTiO_3$ nanoparticles on electrooptic performance of epoxy-resin-based PDLC devices. *Liq. Cryst.*, 48, 23–34. doi: 10.1080/02678292.2020.1762129

[33] S. Al-Zangana, M. Turner, and I. Dierking (2017). A comparison between size dependent paraelectric and ferroelectric $BaTiO_3$ nanoparticle doped nematic and ferroelectric liquid crystals. *J. Appl. Phys.*, 121, 085105.

[34] Yu. Garbovskiy and A. Glushchenko (2017). Ferroelectric nanoparticles in liquid crystals: recent progress and current challenges. *Nanomaterials*, 7, 361. doi:10.3390/nano7110361

[35] A. Klug and D. L. D. Caspar (1961). The structure of small viruses, *Adv. Virus Research*, 7, 225–325.

[36] F. C. Bawden, N. W. Pirie, J. D. Bernal, and I. Fankuchen (1936). Liquid crystalline substances from virus-infected plants. *Nature*, 138, 1051–1052.

[37] W. M. Stanley (1935). Isolation of a crystalline protein possessing the properties of tobacco-mosaic virus. *Science*, 81, 644–645.

[38] M. A. Lauffer (1944). The size and shape of tobacco mosaic virus particles. *J. Am. Chem. Soc.*, 66, 1188–1194.

[39] R. Oldenbourg, X. Wen, R. B. Meyer, and D. L. D. Caspar (1988). Orientational distribution function in nematic tobacco-mosaic-virus liquid crystals measured by X-ray diffraction. *Phys. Rev. Lett.*, 61, 1851–1854.

[40] S. Fraden, G. Maret, and D. L. D. Caspar (1993). Angular correlations and the isotropic-nematic phase transition in suspensions of tobacco mosaic virus. *Phys. Rev. E*, 48, 2816–2837.

[41] X. Wen, R. B. Meyer, and D. L. D. Caspar (1989). Observation of smectic-a ordering in a solution of rigid-rod-like particles. *Phys. Rev. Lett.*, 63, 2760–2763.

[42] M. Hirai, S. Arai, T. Takizawa, Y. Yabuki, and Y. Sano (1997). Dynamics and phase behavior of a supermacromolecular suspension under a magnetic field studied by time-resolved X-ray scattering. *Phys. Rev. B*, 55, 3490–3496.

[43] H. Graf and H. Löwen (1999). Phase diagram of tobacco mosaic virus solutions. *Phys. Rev. E*, 59, 1932–1942.

[44] Z. Dogic and S. Fraden (2000). Cholesteric phase in virus suspensions. *Langmuir*, 16, 7820–7824.

[45] Z. Dogic and S. Fraden (1997). Smectic phase in a colloidal suspension of semiflexible virus particles. *Phys. Rev. Lett*, 78, 2417–2420.

[46] Z. Dogic and S. Fraden (2006). Ordered phases of filamentous viruses. *Curr. Opinion Colloid Interface Sci.*, 11, 47–55.

[47] Z. Dogic (2016). Filamentous phages as a model system in soft matter physics. *Frontiers in Microbiology*, 7, 1013. doi: 10.3389/fmicb.2016.01013

[48] D. Trache, A. F. Tarchoun, M. Derradji, T. S. Hamidon, N. Masruchin, N. Brosse, and M. H. Hussin (2020). Nanocellulose: From fundamentals to advanced applications. *Front. Chem.*, 8, 392. doi: 10.3389/fchem.2020.00392

[49] Y. Habibi, L. A. Lucia, and O. J. Rojas (2010). Cellulose nanocrystals: Chemistry, self-assembly, and applications. *Chem. Rev.*, 110, 3479–3500.

[50] G. Johnsy and S. N. Sabapathi (2015). Cellulose nanocrystals: Synthesis, functional properties, and applications. *Nanotech. Sci. Appl.*, 8, 45–54.

[51] X. M. Dong, T. Kimura, J.-F. Revol, and D. G. Gray, Effects of ionic strength on the isotropic-chiral nematic phase transition of suspensions of cellulose crystallites. *Langmuir*, 12, 2076–2082.

[52] M. Pääkkö, M. Ankerfors, H. Kosonen, A. Nykänen, S. Ahola, M. Österberg, J. Ruokolainen, J. Laine, P. T. Larsson, O. Ikkala, and T. Lindström (2007). Enzymatic hydrolysis combined with mechanical shearing and high-pressure homogenization for nanoscale cellulose fibrils and strong gels. *Biomacromolecules*, 8, 1934–1941.

[53] S. Beck-Candanedo, M. Roman, and D. G. Gray (2005). Effect of reaction conditions on the properties and behavior of wood cellulose nanocrystal suspensions. *Biomacromolecules*, 6, 1048–1054.

[54] C. Honorato-Rios, A. Kuhnhold, J. R. Bruckner, R. Dannert, T. Schilling, and J. P. F. Lagerwall (2016). Equilibrium liquid crystal phase diagrams and detection of kinetic arrest in cellulose nanocrystal suspensions. *Front. Mater.*, 3, 21. doi: 10.3389/fmats.2016.00021

[55] E. E. Ureña-Benavides, G. Ao, V. A. Davis, C. L. Kitchens (2011). Rheology and phase behavior of lyotropic cellulose nanocrystal suspensions. *Macromolecules*, 44, 8990–8998.

[56] C. Schütz, M. Agthe, A. B. Fall, K. Gordeyeva, V. Guccini, M. Salajková, T. S. Plivelic, J. P. F. Lagerwall, G. Salazar-Alvarez, and L. Bergström (2015). Rod packing in chiral nematic cellulose nanocrystal dispersions studied by small-angle X-ray scattering and laser diffraction. *Langmuir*, 31, 6507–6513.

[57] G. Nyström, M. Arcari, J. Adamcik, I. Usov, and R. Mezzenga (2018). Nanocellulose fragmentation mechanisms and inversion of chirality from the single particle to the cholesteric phase. *ACS Nano*, 12, 5141–5148.

[58] C. Duan, Z. Cheng, B. Wang, J. Zeng, J. Xu, J. Li, W. Gao, and K. Chen (2021). Chiral photonic liquid crystal films derived from cellulose nanocrystals. *Small*, 17, 2007306.

[59] T. G. Parton, R. M. Parker, G. T. van de Kerkhof, A. Narkevicius, J. S. Haataja, B. Frka-Petesic, and S. Vignolini (2022). Chiral self-assembly of cellulose nanocrystals is driven by crystallite bundles. *Nature Communications*, 13, 2657. https://doi.org/10.1038/s41467-022-30226-6

[60] S. N. Fernandes, P. L. Almeida, N. Monge, L. E. Aguirre, D. Reis, C. L. P. de Oliveira, A. M. F. Neto, P. Pieranski, and M. H. Godinho (2017). Mind the microgap in iridescent cellulose nanocrystal films. *Adv. Mater.*, 29, 1603560.

[61] Q. Wu, Y. Meng, K. Conchab, S. Wang, Y. Li, L. Ma, and S. Fu (2013). Influence of temperature and humidity on nano-mechanical properties of cellulose nanocrystal films made from switchgrass and cotton. *Industrial Crops and Products*, 48, 28–35.

[62] J. H. Park, J. H. Noh, C. Schütz, G. Salazar-Alvarez, G. Scalia, L. Bergström, and J. P. F. Lagerwall (2014). Macroscopic control of helix orientation in films dried from cholesteric liquid-crystalline cellulose nanocrystal suspensions. *ChemPhysChem*, 15, 1477–1484.

[63] Q. Wu, Y. Meng, S. Wang, Y. Li, S. Fu, L. Ma, and D. Harper (2014). Rheological behavior of cellulose nanocrystal suspension: influence of concentration and aspect ratio. *J. Appl. Polym. Sci.*, 131, 40525. doi: 10.1002/APP.40525

[64] D. Liu, X. Chen, Y. Yue, M. Chena, and Q. Wu (2011). Structure and rheology of nanocrystalline cellulose. *Carbohydrate Polymers*, 84, 316–322.

[65] S. Shafiei-Sabet, W. Y. Hamad, and S. G. Hatzikiriakos (2014). Ionic strength effects on the microstructure and shear rheology of cellulose nanocrystal suspensions. *Cellulose*, 21, 3347–3359. doi: 10.1007/s10570-014-0407-z

[66] A. Querejeta-Fernández, G. Chauve, M. Methot, J. Bouchard, and E. Kumacheva (2014). Chiral plasmonic films formed by gold nanorods and cellulose nanocrystals, *J. Am. Chem. Soc.*, 136, 4788–4793.

[67] Z. Cheng, Y. Ma, L. Yang, F. Cheng, Z. Huang, A. Natan, H. Li, Y. Chen, D. Cao, Z. Huang, Y. Wang, Y. Liu, R. Yang, and H. Zhu (2019). Plasmonic-enhanced cholesteric films: Coassembling anisotropic gold nanorods with cellulose nanocrystals. *Adv. Optical Mater.*, 7, 1801816.

[68] Q. Liu, M. G. Campbell, J. S. Evans, and I. I. Smalyukh (2014). Orientationally ordered colloidal co-dispersions of gold nanorods and cellulose nanocrystals. *Adv. Mater.*, 26, 7178–7184.

[69] G. Chu, X. Wang, T. Chen, J. Gao, F. Gai, Y. Wang, and Y. Xu (2015). Optically tunable chiral plasmonic guest–host cellulose films weaved with long-range ordered silver nanowires. *ACS Appl. Mater. Interfaces*, 7, 11863–11870.

[70] B. Frka-Petesic, H. Radavidson, B. Jean, and L. Heux (2017). Dynamically controlled iridescence of cholesteric cellulose nanocrystal suspensions using electric fields. *Adv. Mater.*, 29, 1606208.

[71] K. J. De France, K. G. Yager, T. Hoare, and E. D. Cranston (2016). Cooperative ordering and kinetics of cellulose nanocrystal alignment in a magnetic field. *Langmuir*, 32, 7564–7571.

[72] J. P. F. Lagerwall, C. Schütz, M. Salajkova, J. Noh, J. H. Park, G. Scalia, and L. Bergström (2014). Cellulose nanocrystal-based materials: From liquid crystal self-assembly and glass formation to multifunctional thin films. *NPG Asia Materials*, 6, e80. doi:10.1038/am.2013.69

[73] D. G. Gray and X. Mu (2015). Chiral nematic structure of cellulose nanocrystal suspensions and films; polarized light and atomic force microscopy. *Materials*, 8, 7873–7888. doi:10.3390/ma8115427

[74] D. G. Gray (2016). Recent advances in chiral nematic structure and iridescent color of cellulose nanocrystal films. *Nanomaterials*, 6, 213. doi:10.3390/nano6110213

[75] Y, Xu, A. Atrens, and J. R. Stokes (2020). A review of nanocrystalline cellulose suspensions: Rheology, liquid crystal ordering and colloidal phase behaviour. *Adv. Colloid Interf. Sci.*, 275, 102076.

[76] R. Kádár, S. Spirk, and T. Nypelö (2021). Cellulose nanocrystal liquid crystal phases: Progress and challenges in characterization using rheology coupled to optics, scattering, and spectroscopy. *ACS Nano*, 15, 7931–7945.

[77] A. A. Moud and A. A. Moud (2022). Cellulose nanocrystals (CNC) liquid crystalline state in suspension: An overview. *Appl. Biosci.*, 1, 244–278. https://doi.org/10.3390/applbiosci1030016

[78] S. Sridevi, S. K. Prasad, G. G. Nair, V. D'Britto, and B. L. V. Prasad (2010). Enhancement of anisotropic conductivity, elastic, and dielectric constants in a liquid crystal-gold nanorod system. *Appl. Phys. Lett.*, 97, 151913. https://doi.org/10.1063/1.3499744

[79] S. Khatua, W.-S. Chang, P. Swanglap, J. Olson, and S. Link (2011). Active modulation of nanorod plasmons. *Nano Lett.*, 11, 3797–3802.

[80] Z. Su, J. Yin, and X. Zhao (2015). Soft and broadband infrared metamaterial absorber based on gold nanorod/liquid crystal hybrid with tunable total absorption. *Scientific Reports*, 5, 16698. doi: 10.1038/srep16698

[81] U. Tritschler, I. Zlotnikov, P. Keckeis, H. Schlaad, and H. Cölfen (2014). Optical properties of self-organized gold nanorod–polymer hybrid films. *Langmuir*, 30, 13781–13790.

[82] Y. Wang, A. Dang, Z. Zhang, R. Yin, Y. Gao, L. Feng, and S. Yang (2020). Repeatable and reprogrammable shape morphing from photoresponsive gold nanorod/liquid crystal elastomers. *Adv. Mater.*, 32, 2004270.

[83] X. Yanga, Y. Chena, X. Zhanga, P. Xuea, P. Lva, Y. Yanga, L. Wanga, and W. Feng (2022). Bioinspired light-fueled water-walking soft robots based on liquid crystal network actuators with polymerizable miniaturized gold nanorods. *Nano Today*, 43, 101419.

[84] B. Rožič, J. Fresnais, C. Molinaro, J. Calixte, S. Umadevi, S. Lau-Truong, N. Felidj, T. Kraus, F. Charra, V. Dupuis, T. Hegmann, C. Fiorini-Debuisschert, B. Gallas, and E. Lacaze (2017). Oriented gold nanorods and gold nanorod chains within smectic liquid crystal topological defects. *ACS Nano*, 11, 6728–6738.

[85] S. Umadevi, X. Feng, and T. Hegmann (2013). Large area self-assembly of nematic liquid-crystal-functionalized gold nanorods. *Adv. Funct. Mater.*, 23, 1393–1403.

[86] W.-K. Fong, T. L. Hanley, B. Thierry, N. Kirby, L. J. Waddington, and B. J. Boyd (2012). Controlling the nanostructure of gold nanorod–lyotropic liquid-crystalline hybrid materials using near-infrared laser irradiation. *Langmuir*, 28, 14450–14460.

[87] X. Liu, G. Qi, A. M. G. Park, A. Rodriguez-Gonzalez, A. Enotiadis, W. Pan, V. Kosma, G. D. Fuchs, B. J. Kirby, and E. P. Giannelis (2019). Scalable synthesis of switchable assemblies of gold nanorod lyotropic liquid crystal nanocomposites. *Small*, 15, 1901666.

[88] D. Duc, P. R. Stoddart, S. L. McArthur, R. M. I. Kapsa, A. F. Quigley, M. Boyd-Moss, and S. E. Moulton (2019). Fabrication of a biocompatible liquid crystal graphene oxide–gold nanorods electro- and photoactive interface for cell stimulation. *Adv. Healthcare Mater.*, 8, 1801321.

[89] S. N. Hosseini, X. Chen, P. J. Baesjou, A. Imhof, and A. van Blaaderen (2022). Synthesis and characterization of anatase TiO2 nanorods: Insights from nanorods' formation and self-assembly. *Appl. Sci.*, 12, 1614. https://doi.org/10.3390/app12031614

[90] L. Liu, Y. Bai, F. Wang, and N. Liu (2012). Fabrication and characterizes of TiO_2 nanomaterials templated by lyotropic liquid crystal. *Advanced Materials Research*, 399-401, 532–537. https://doi.org/10.4028/www.scientific.net/AMR.399-401.532

[91] Z. Ren, C. Chen, R. Hu, K. Mai, G. Qian, and Z. Wang (2012). Two-step self-assembly and lyotropic liquid crystal behavior of TiO_2 nanorods. *Journal of Nanomaterials*, 180989. doi:10.1155/2012/180989

[92] S. N. Hosseini, A. Grau-Carbonell, A. G. Nikolaenkova, X. Xie, X. Chen, A. Imhof, A. van Blaaderen, and P. J. Baesjou (2020). Smectic liquid crystalline titanium dioxide nanorods: reducing attractions by optimizing ligand density. *Adv. Funct. Mater.*, 30, 2005491.

[93] X. Peng, L. Manna, W. Yang, J. Wickham, E. Scher, A. Kadavanich, and A. P. Alivisatos (2000). Shape control of CdSe nanocrystals. *Nature*, 404, 59–61.

[94] J. Hu, L. Li, W. Yang, L. Manna, L. Wang, and A. P. Alivisatos (2001). Linearly polarized emission from colloidal semiconductor quantum rods. *Science*, 292, 2060–2063.

[95] L. Li, J. Walda, L. Manna, and A. P. Alivisatos (2002). Semiconductor nanorod liquid crystals. *Nano Letters*, 2, 557–560.

[96] F. Kim, S. Kwan, J. Akana, and P. Yang (2001). Langmuir–Blodgett nanorod assembly. *J. Am. Chem. Soc.*, 123, 4360–4361.

[97] L.-S. Li and A. P. Alivisatos (2003). Semiconductor nanorod liquid crystals and their assembly on a substrate. *Adv. Mater.*, 15, 408–411.

[98] K. A. Stancil (2014). Evaporation assisted CdSe nanorod assembly by small angle X-ray scattering and Langmuir adsorption. *J. Phys. Chem. B*, 118, 14695–14702.

[99] T. Xu and V. A. Davis (2014). Liquid crystalline phase behavior of silica nanorods in dimethyl sulfoxide and water. *Langmuir*, 30, 4806–4813.

[100] M. D. Lynch and D. L. Patrick (2002). Organizing carbon nanotubes with liquid crystals. *Nano Letters*, 2, 1197–1201.

[101] I. Dierking, G. Scalia, P. Morales, and D. LeClere (2004). Aligning and reorienting carbon nanotubes with nematic liquid crystals. *Adv. Mater.*, 16, 865–860.

[102] I. Dierking, G. Scalia, and P. Morales (2005). Liquid crystal–carbon nanotube dispersions, *J. App. Phys.*, 97, 044309.

[103] I. Dierking and S. E. San (2005). Magnetically steered liquid crystal–nanotube switch. *Appl. Phys. Lett.*, 87, 233507.

[104] I. Dierking, K. Casson, and R. Hampson (2008). Reorientation dynamics of liquid crystal–nanotube dispersions. *Jpn. J. Appl. Phys.*, 47, 6390–6393.

[105] J. P. F. Lagerwall, G. Scalia, M. Haluska, U. Dettlaff-Weglikowska, F. Giesselmann, and S. Roth (2006). Simultaneous alignment and dispersion of carbon

nanotubes with lyotropic liquid crystals. *Phys. Stat. Sol. (b)*, 243, 3046–3049. doi:10.1002/pssb.200669146.

[106] J. Lagerwall, G. Scalia, M. Haluska, U. Dettlaff-Weglikowska, S. Roth, and F. Giesselmann (2007). Nanotube alignment using lyotropic liquid crystals. *Adv. Mater.*, 19, 359–364.

[107] S. Schymura, E. Enz, S. Roth, G. Scalia, and J. P. F. Lagerwall (2009). Macroscopic-scale carbon nanotube alignment via self-assembly in lyotropic liquid crystals, *Synthetic Metals*, 159, 2177–2179.

[108] G. Scalia, C. von Bühler, C. Hägele, S. Roth, F. Giesselmann, and J. P. F. Lagerwall (2008). Spontaneous macroscopic carbon nanotube alignment via colloidal suspension in hexagonal columnar lyotropic liquid crystals. *Soft Matter*, 4, 570–576.

[109] V. Weiss, R. Thiruvengadathan, and O. Regev (2006). Preparation and characterization of a carbon nanotube–lyotropic liquid crystal composite. *Langmuir*, 22, 854–856.

[110] W. Jiang, B. Yu, W. Liu, and J. Hao (2007). Carbon nanotubes incorporated within lyotropic hexagonal liquid crystal formed in room-temperature ionic liquids. *Langmuir*, 23, 8549–8553.

[111] W. Song, I. A. Kinloch, and A. H. Windle (2003). Nematic liquid crystallinity of multiwall carbon nanotubes. *Science*, 302, 1363.

[112] S. Zhang, I. A. Kinloch, and A. H. Windle (2006). Mesogenicity drives fractionation in lyotropic aqueous suspensions of multiwall carbon nanotubes. *Nano Letters*, 6, 568–572.

[113] S. Badaire, C. Zakri, M. Maugey, A. Derre, J. N. Barisci, G. Wallace, and P. Poulin (2005). Liquid crystals of dna-stabilized carbon nanotubes. *Adv. Mater.*, 17, 1673–1676.

[114] N. Puech, C. Blanc, E. Grelet, C. Zamora-Ledezma, M. Maugey, C. Zakri, E. Anglaret, and P. Poulin (2011). Highly ordered carbon nanotube nematic liquid crystals. *J. Phys. Chem. C*, 115, 3272–3278.

[115] N. Puech, E. Grelet, P. Poulin, C. Blanc, and P. van der Schoot (2010). Nematic droplets in aqueous dispersions of carbon nanotubes. *Phys. Rev. E*, 82, 020702(R).

[116] G. Ao, D. Nepal, M. Aono, and V. A. Davis (2011). Cholesteric and nematic liquid crystalline phase behavior of double-stranded dna stabilized single-walled carbon nanotube dispersions. *ACS Nano*, 5, 1450–1458.

[117] J. P. F. Lagerwall and G. Scalia (2008). Carbon nanotubes in liquid crystals. *J. Mater. Chem.*, 18, 2890–2898.

[118] G. Scalia (2010). Alignment of carbon nanotubes in thermotropic and lyotropic liquid crystals. *ChemPhysChem*, 11, 333 – 340.

[119] M. Bravo-Sanchez, T. J. Simmons, and M. A. Vidal (2010). Liquid crystal behavior of single wall carbon nanotubes. *Carbon*, 48, 3531–3542.

[120] V. A. Davis (2011). Liquid crystalline assembly of nanocylinders. *J. Mater. Res.*, 26, 140–153.

[121] S. Zhang, P. W. Majewski, G. Keskar, L. D. Pfefferle, C. O. Osuji (2011). Lyotropic self-assembly of high-aspect-ratio semiconductor nanowires of single-crystal ZnO. *Langmuir*, 27, 11616–11621.

[122] S. Zhang, C. I. Pelligra, G. Keskar, P. W. Majewski, F. Ren, L. D. Pfefferle, C. O. Osuji (2011). Liquid crystalline order and magnetocrystalline anisotropy in magnetically doped semiconducting ZnO nanowires. *ACS Nano*, 5, 8357–8364.

[123] S. Murali, T. Xu, B. D. Marshall, M. J. Kayatin, K. Pizarro, V. K. Radhakrishnan, D. Nepal, and V. A. Davis (2010). Lyotropic liquid crystalline self-assembly in dispersions of silver nanowires and nanoparticles. *Langmuir*, 26, 11176–11183.

[124] T. Xu and V. A. Davis (2015). Rheology and shear-induced textures of silver nanowire lyotropic liquid crystals. *Journal of Nanomaterials*, 939587. http://doi.org/10.1155/2015/939587.

[125] S. J. Kim, H. H. An, S. J. Lee, J. H. Lee, Y. H. Kim, C. S. Yoon, and S. H. Suh (2012). Formation of Ag nanostrings induced by lyotropic liquid_crystalline phospholipid multilayer. *Langmuir*, 28, 259–263.

[126] C. Zhi, Y. Bando, W. Wang, C. Tang, H. Kuwahara, and D. Golberg (2007). DNA-mediated assembly of boron nitride nanotubes. *Chem. Asian J.*, 2, 1581–1585.

[127] C. J. Simonsen Ginestra, C. Martínez-Jiménez, A. Matatyaho Ya'akobi, O. S. Dewey, A. D. Smith McWilliams, R. J. Headrick, J. A. Acapulco, L. R. Scammell, M. W. Smith, D. V. Kosynkin, D. M. Marincel, C. Park, S.-H. Chu, Y. Talmon, A. A. Martí, and M. Pasquali (2022). Liquid crystals of neat boron nitride nanotubes and their assembly into ordered macroscopic materials. *Nature Communications*, 13, 3136. https://doi.org/10.1038/s41467-022-30378-5.

[128] H. Lim, Y.-K. Kim, H.-S. Kim, T. Lee, M. Monir Hossain, H.-O. Jeong, H. S. Lee, H. Cho, Y. Joo, S. S. Lee, S. Park, H. Rho, H. S. Jeong, M. J. Kim, S. Ahn, S. Y. Moon, K. S. Kim, S. Q. Choi, B. J. Kim, and S. G. Jang (2023). Lyotropic boron nitride nanotube liquid crystals: preparation, characterization, and wet-spinning for fabrication of composite fiber. *ACS Appl. Mater. Interfaces*, 15, 24681–24692.

[129] A. B. Shivanandareddy, M. Kumar, and S. Kumar (2018). Lyotropic liquid crystals of CdS nanoribbons. *J. Mol. Liq.*, 264, 352–357.

[130] E. Paineau, M.-E. M. Krapf, M.-S. Amara, N. V. Matskova, I. Dozov, S. Rouziere, A. Thill, P. Launois, and P. Davidson (2016). A liquid-crystalline hexagonal columnar phase in highly-dilute suspensions of imogolite nanotubes. *Nature Communications*, 7, 10271. doi:10.1038/ncomms10271.

[131] Y. Li, J. Wan, and Z. Gu (2000). The formation of cadmium sulfide nanowires in different liquid crystal systems. *Materials Science and Engineering*, A286, 106–109.

[132] G. N. Karanikolos, N.-L. Law, R. Mallory, A. Petrou, P. Alexandridis, and T. J. Mountziaris (2006). Water-based synthesis of ZnSe nanostructures using amphiphilic block copolymer stabilized lyotropic liquid crystals as templates. *Nanotechnology*, 17, 3121–3128.

[133] Y. Zhu, X. Guo, J. Jin, Y. Shen, X. Guo, and W. Ding (2007). Controllable synthesis of CuS nanotubes and nanobelts using lyotropic liquid crystal templates. *J. Mater. Sci.*, 42, 1042–1045.

[134] Y. Zhu, X. Guo, M. Mo, X. Guo, W. Ding, and Y. Chen (2008). In situ synthesis of horizontally aligned metal–boron alloy nanotubes on a silicon substrate with liquid crystal template. *Nanotechnology*, 19, 405602.

[135] T. Kijima, T. Yoshimura, M. Uota, T. Ikeda, D. Fujikawa, S. Mouri, and S. Uoyama (2004). Noble-metal nanotubes (Pt, Pd, Ag) from lyotropic mixed-surfactant liquid-crystal templates. *Angew. Chem. Int. Ed.*, 43, 228 –232.

[136] A. Takai, Y. Yamauchi, and K. Kuroda Facile formation of single crystalline Pt nanowires on a substrate utilising lyotropic liquid crystals consisting of cationic surfactants. *J. Mater. Chem.*, 19, 4205–4210.

[137] Y. Zhong, C.-L. Xu, L.-B. Kong, and H.-L. Li (2008). Synthesis and high catalytic properties of mesoporous Pt nanowire array by novel conjunct template method. *Applied Surface Science*, 255, 3388–3393.

[138] C. Wang, D. Chen, and X. Jiao (2009). Lyotropic liquid crystal directed synthesis of nanostructured materials. *Sci. Technol. Adv. Mater.*, 10, 023001.

[139] K. S. Novoselov, A. K. Geim, S. V. Morozov, D. Jiang, Y. Zhang, S. V. Dubonos, I. V. Grigorieva, and A. A. Firsov (2004). Electric field effect in atomically thin carbon films. *Science*, 306, 666–668.

[140] N. Behabtu, J. R. Lomeda, M. J. Green, A. L. Higginbotham, A. Sinitskii, D. V. Kosynkin, D. Tsentalovich, A. Nicholas, G. Parra-Vasquez, J. Schmidt, E. Kesselman, Y. Cohen, Y. Talmon, J. M. Tour, and M. Pasquali (2010). Spontaneous high-concentration dispersions and liquid crystals of graphene. *Nature Technology*, 5, 406–411.

[141] J. E. Kim, T. H. Han, S. H. Lee, J. Y. Kim, C. W. Ahn, J. M. Yun, and S. O. Kim (2011). Graphene oxide liquid crystals. *Angew. Chem. Int. Ed.*, 50, 3043–3047.

[142] Z. Xu and C. Gao (2011). Aqueous liquid crystals of graphene oxide. *ACS Nano*, 5, 2908–2915.

[143] F. Lin, X. Tong, Y. Wang, J. Bao, and Z. M. Wang (2015). Graphene oxide liquid crystals: synthesis, phase transition, rheological property, and applications in optoelectronics and display. *Nanoscale Research Letters*, 10, 435.

[144] S. P. Sasikala, J. Lim, I. H. Kim, H. J. Jung, T. Yun, T. H. Han, and S. O. Kim (2018). Graphene oxide liquid crystals: a frontier 2D soft material for graphene-based functional materials. *Chem. Soc. Rev.*, 47, 6013–6045.

[145] R. Jalili, S. H. Aboutalebi, D. Esrafilzadeh, K. Konstantinov, S. E. Moulton, J. M. Razal, and G. G. Wallace (2013). Organic solvent-based graphene oxide liquid crystals: a facile route toward the next generation of self-assembled layer-by-layer multifunctional 3D architectures. *ACS Nano*, 7, 3981–3990.

[146] R. T. M. Ahmad, S.-H. Hong, T.-Z. Shen, A. R. Masud, and J.-K. Song (2016). Effect of solvents on the electro-optical switching of graphene oxide dispersions. *Appl. Phys. Lett.*, 108, 251903.

[147] R. T. M. Ahmad, S.-H. Hong, T.-Z. Shen, and J.-K. Song (2015). Optimization of particle size for high birefringence and fast switching time in electrooptical switching of graphene oxide dispersions. *Optics Express*, 23, 4436. doi:10.1364/OE.23.004435.

[148] S. H. Aboutalebi, M. M. Gudarzi, Q. B. Zheng, and J.-K. Kim (2011). Spontaneous formation of liquid crystals in ultralarge graphene oxide dispersions. *Adv. Funct. Mater.*, 21, 2978–2988.

[149] B. Dan, N. Behabtu, A. Martinez, J. S. Evans, D. V. Kosynkin, J. M. Tour, M. Pasquali, and I. I. Smalyukh (2011). Liquid crystals of aqueous, giant graphene oxide flakes. *Soft Matter*, 7, 11154–11159.

[150] S. Shahini, S. Campidelli, and G. Scalia (2019). Large graphene oxide flakes: From isotropic to nematic liquid crystal phase transition. *Phys. Stat. Sol. B*, 256, 1900411.

[151] B. Konkena and S. Vasudevan (2014). Glass, gel, and liquid crystals: Arrested states of graphene oxide aqueous dispersions. *J. Phys. Chem. C*, 118, 21706–21713.

[152] T.-Z. Shen, S.-H. Hong, and J.-K. Song (2014). Electro-optical switching of graphene oxide liquid crystals with an extremely large Kerr coefficient. *Nature Materials*, 13, 394–399.

[153] S. Al-Zangana, M. Iliut, M. Turner, A. Vijayaraghavan, and I. Dierking (2017). Confinement effects on lyotropic nematic liquid crystal phases of graphene oxide dispersions. *2D Mater.*, 4, 041004.

[154] R. Jalili, S. H. Aboutalebi, D. Esrafilzadeh, R. L. Shepherd, J. Chen, S. Aminorroaya-Yamini, K. Konstantinov, A. I. Minett, J. M. Razal, and G. G. Wallace (2013). Scalable one-step wet-spinning of graphene fibers and yarns from liquid crystalline dispersions of graphene oxide: Towards multifunctional textiles. *Adv. Funct. Mater.*, 23, 5345–5354.

[155] Z. Xu and C. Gao (2011). Graphene chiral liquid crystals and macroscopic assembled fibres. *Nature Communications*, 2, 571.

[156] Z. Xu, H. Sun, X. Zhao, and C. Gao (2013). Ultrastrong fibers assembled from giant graphene oxide sheets. *Adv. Mater.*, 25, 188–193.

[157] Z. Liu, Z. Xu, X. Hu, and C. Gao (2013). Lyotropic liquid crystal of polyacrylonitrile-grafted graphene oxide and its assembled continuous strong nacre-mimetic fibers. *Macromolecules*, 46, 6931–6941.

[158] Z. Xu and C. Gao (2014). Graphene in macroscopic order: Liquid crystals and wet-spun fibers. *Acc. Chem. Res.*, 47, 1267–1276.

[159] F. Meng, W. Lu, Q. Li, J.-H. Byun, Y. Oh, and T.-W. Chou (2015). Graphene-based fibers: A review. *Adv. Mater.*, 27, 5113–5131.

[160] R. Narayan, J. E. Kim, J. Y. Kim, K. E. Lee, and S. O. Kim (2016). Graphene oxide liquid crystals: Discovery, evolution and applications. *Adv. Mater.*, 28, 3045–3068.

[161] A. P. Draude and I. Dierking (2019). Lyotropic liquid crystals from colloidal suspensions of graphene oxide. *Crystals*, 9, 455. doi:10.3390/cryst9090455.

[162] R. Jalili, S. Aminorroaya-Yamini, T. M. Benedetti, S. H. Aboutalebi, Y. Chao, G. G. Wallace, and D. L. Officer (2016). Processable 2D materials beyond graphene: MoS_2 liquid crystals and fibres. *Nanoscale*, 8, 16862–16867.

[163] M. Naguib, M. Kurtoglu, V. Presser, J. Lu, J. Niu, M. Heon, L. Hultman, Yu. Gogotsi, and M. W. Barsoum (2011). Two-dimensional nanocrystals produced by exfoliation of Ti_3AlC_2. *Adv. Mater.*, 23, 4248–4253.

[164] Yu. Gogotsi and B. Anasori (2019). The rise of MXenes. *ACS Nano*, 13, 8491–8494.

[165] J. Zhang, N. Kong, S. Uzun, A. Levitt, S. Seyedin, P. A. Lynch, S. Qin, M. Han, W. Yang, J. Liu, X. Wang, Yu. Gogotsi, and J. M. Razal (2020). Scalable manufacturing

of free-standing, strong $Ti_3C_2T_x$ MXene films with outstanding conductivity. *Adv. Mater.*, 32, 2001093.

[166] J. Zhang, S. Uzun, S. Seyedin, P. A. Lynch, B. Akuzum, Z. Wang, S. Qin, M. Alhabeb, C. E. Shuck, W. Lei, E. C. Kumbur, W. Yang, X. Wang, G. Dion, J. M. Razal, and Yu. Gogotsi (2020). Additive-free MXene liquid crystals and fibers. *ACS Cent. Sci.*, 6, 254–265.

[167] K. Aldren, S. Usman, J. Zhang, S. Qin, Y. Yao, P. A. Lynch, P. Mota-Santiago, M. Naebe, L. C. Henderson, D. Hegh, and J. M. Razal (2022). Inducing liquid crystallinity in dilute MXene dispersions for facile processing of multifunctional fibers. *J. Mater. Chem. A*, 10, 4770–4781.

[168] Y. Xia, T. S. Mathis, M.-Q. Zhao, B. Anasori, A. Dang, Z. Zhou, H. Cho, Yu. Gogotsi, and S. Yang (2018). Thickness-independent capacitance of vertically aligned liquid-crystalline MXenes, *Nature*, 557, 409–414.

[169] M. Naguib (2020). Multifunctional pure MXene fiber from liquid crystals of only water and MXene. *ACS Cent. Sci.*, 6, 344–346.

[170] S. Seyedin, E. Roswell, S. Yanza, and J. M. Razal (2017). Knittable energy storing fiber with high volumetric performance made from predominantly MXene nanosheets. *J. Mater. Chem. A*, 5, 24076–24082.

11
Applications

11.1 Introduction

As emphasized throughout this book, some of the characteristics of mixtures that present lyotropic liquid crystalline mesophases give them unique properties. Among these characteristics we can stress: the structures of the mesophases; phase transitions as a function of temperature and relative concentrations of the components; existence of direct and inverted phases; in some structures one may observe large area to volume ratios; nanostructures with different electrical properties in the same basic unit.

In this chapter we discuss some applications of lyotropics in the food, cosmetic, pharmaceutical, and chemical industries and we will finish with some applications in nanotechnology.

11.2 Traditional applications in the food and cosmetics industry

Different aspects can be considered in the application of lyotropics in the food industry. Among them, we can consider taste and aroma, consistence sensation of the product, oxidation retardation or prevention, and locus in the digestive system to deliver substances [1]. Concerning this last aspect, the release of particular substances was studied by measuring the diffusion coefficient and its relationship with the dimensionality of the lipid structures, where the substances are in contact [2].

The permeation of the substances (e.g., caffeine, limonene, glucose) throughout the amphiphilic membranes depends not only on the type of amphiphile (e.g., monolinoleine, phosphatidylcholine), but also on the structure (e.g., bicontinuous, lamellar, micellar, direct and inverted). Depending on the electric characteristics of the target substances present in foods, direct or inverted lyotropic structures can be engineered.

The presence of lyotropic structures in foods may control the reaction that confers flavour to browed foods. This reaction between amino acids and sugars is known as Maillard's reaction [3]. This control occurs due to the permeability of certain substances across the amphiphilic bilayer of the lyotropic structure, which has the ability to maintain, for a certain time, molecules separated in the bulk. This physical separation of molecules, also, controls oxidative reactions.

Lyotropic Liquid Crystals. Ingo Dierking and Antônio Martins Figueiredo Neto, Oxford University Press.
 DOI: 10.1093/9780191924927.003.0011

Food supplements encapsulated in amphiphilic lyotropic structural units are widely used, with evident gains in the process of controlled delivery and body-absorption efficiency [4].

Lyotropics are also present in the cosmetics industry. Probably, the most important goal to be achieved by the products of this industry, is to keep the health, flexibility, hydration, and, finally, the appearance of the stratum corneum of the skin [5]. The term 'cosmeceutical' has been employed to merge the two concepts cosmetic and pharmaceutical, to refer to the use of drugs for different purposes, absorbed by the human skin [6]. The basic characteristic of many amphiphilic molecules that make them useful to achieve that task, is their biocompatibility and the formation of nanoscale structures. In these structures, drugs of different electric characteristics (e.g., polar, nonpolar, ionic), and water, may be incorporated. As lyotropics are sensitive to changes in temperature, leading to phase transitions and even complete destruction of the former structures, drugs may be released when the mesophase comes into contact with the skin.

Besides this 'drug delivery' application, lyotropics are also employed as cleansers. One of these functions is makeup removers. Another is the use of a liquid crystal cleansing oil, that has to be mixed with water, and then washed from the skin.

As lyotropics are mixtures of different molecules, the presence of the material to be delivered may alter the structure of the mesophase, determined before the addition of the drug. This aspect is a challenge in industrial cosmetic manipulation since a new topology of the partial phase diagram may occur after the addition of the new component to the mixture.

An interesting example was reported by Yhirayha and co-workers [7], where mulberry stem extract was incorporated in lamellar lyotropics. This substance has anti-tyrosinase and antioxidant activities. The mixtures investigated were: PEG-7 glyceryl cocoate/tridecyl salicylate/water; mixed surfactant/tridecyl salicylate/water; PEG-7 glyceryl cocoate/n-dodecane/water and mixed surfactant/n-dodecane/water. The lyotropic structure, where the active substance was incorporated, is lamellar. In particular, the lamellar phase has structural similarities with the cellular packing of the skin. Mixtures of PEG-7 glyceryl cocoate/n-dodecane/water, where the active substance was incorporated, revealed small modifications on the correlation distance of the lamellar structure, at physiological temperatures and at 40 °C, modifications of this parameter were observed (e.g., the correlation distance of the lamellar structure decreased—Figure 11.1).

Another example is the addition of black *Dimocarpus longan* seed extracts to a lyotropic lamellar structure [6]. The substance, obtained from these fruits, is a phenolic compound with anti-hyaluronidase and antioxidant activities. Lyotropic mixtures composed of 5% w/w Lexfeel® D5 oil, 0.5% w/w Carbopol® U21, 2% w/w Liquid Crystal Cream Maker, and 92.5% w/w DI water, presenting a lamellar structure at room temperature, were used in these studies.

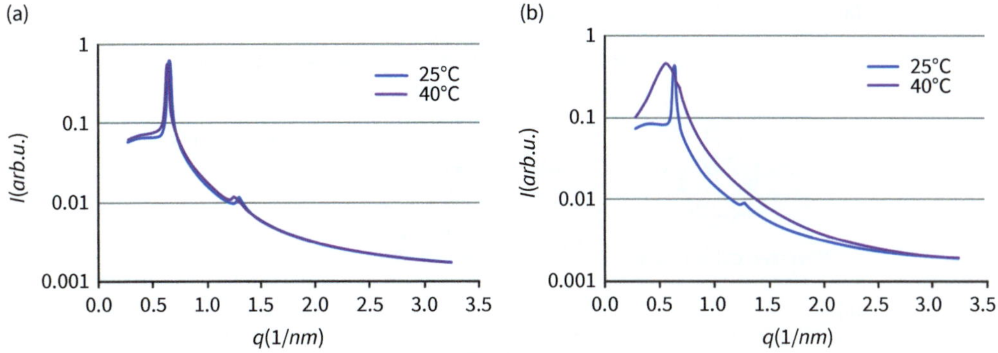

Figure 11.1 Small-angle, X-ray scattering results of the mixture PEG-7 glyceryl cocoate/n-dodecane/water, (a) without and (b) with the active substance. q is the modulus of the scattering vector.

11.3 Drug delivery

This is one of the most interesting applications of lyotropics, with important consequences for our daily life and well-being. The aim of this approach is to achieve a method to increase the concentration of pharmaceuticals in a particular place of the body, and a controlled process of releasing the drug. An important aspect, in some of these applications, is that the encapsulation of drugs in lyotropic structures protects the drug from enzymatic degradation during the time from its administration until it reaches the place of action. Lyotropic structures, e.g., lamellar, reversed cubic, and hexagonal have been employed to keep, transport, and release low molar weight molecules, for different applications [8].

Figure 11.2 shows a sketch of lyotropic structures, with examples of drug incorporation. The main structure is composed of non-toxic and biodegradable polar lipids. Examples of amphiphilic materials employed for this purpose are [9]: glycerol monooleate, phytantriol, glycerol dioleate, oleyl glycerate, and phytanyl glycerate. These structures have regions with different electrical properties, allowing the incorporation of drugs of different electrical affinities. To be useful for drug delivery, nanoparticles of the lyotropic structure have to be produced. These nanoparticles are named cubosomes and hexosomes, when the structure of the particles is cubic or hexagonal, respectively.

There exist four main routes to prepare the lipid nanoparticles: top-down; bottom-up; heat treatment; and spray drying [8]. Figure 11.3 shows the cryo-transmission electron microscopy pictures of inverse cubic (Figure 11.3(a)) and hexagonal (Figure 11.3(b)) [10] nanoparticles. A disadvantage of the use of cubic and hexagonal lyotropic structures in drug delivery is their high viscosity. This drawback could be overcome by using a, less viscous, lyotropic lamellar structure to encapsulate the drug [11,12].

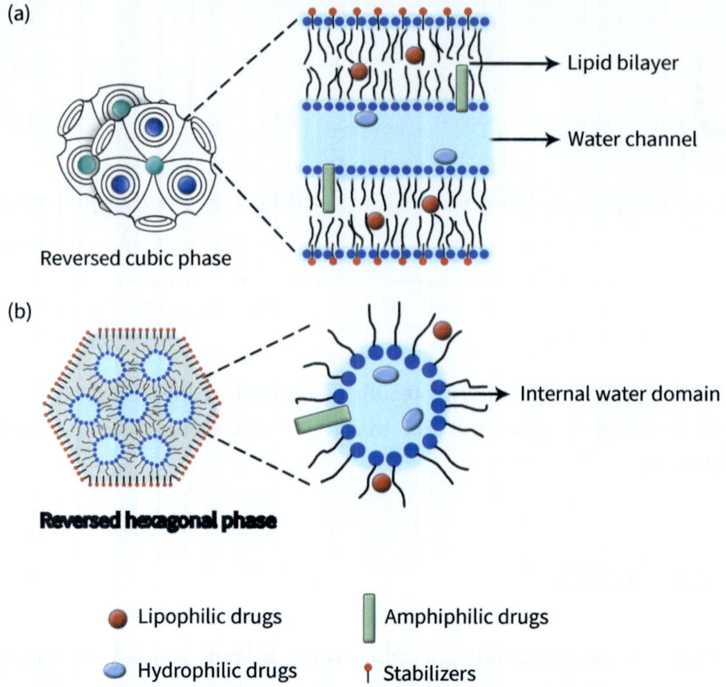

Figure 11.2 (a) Sketch of the lyotropic reversed cubic and (b) reversed hexagonal structures, with lipophilic, hydrophilic, and amphiphilic drug incorporation.
(Reproduced with permission from ref.[8]).

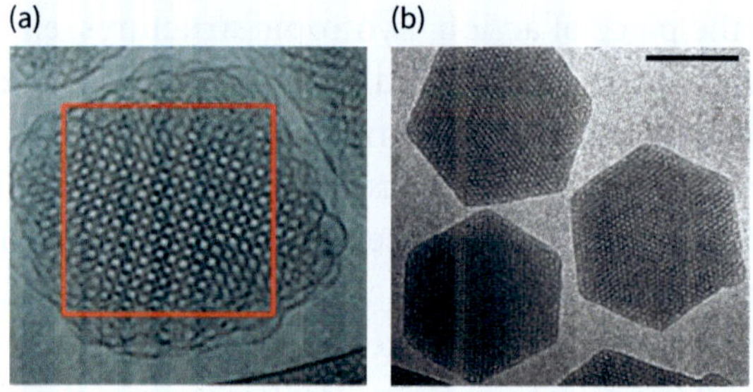

Figure 11.3 Cryo-transmission electron microscopy pictures of the (a) inverse cubic particles in water. In the red square (size about 240 nm), we see an almost perfect cubic structure, monoglyceride/surfactant mixtures in water. Reproduced, with permission, from [10].
(b) inverse hexagonal particles, mixture of diglycerol monooleate/glycerol dioleate 50/50 wt/wt lipid mixture that was dispersed with the help of poloxamer 407 (lipid/poloxamer 407 9/1 wt/wt), scale bar represents 100 nm.
(Reproduced with permission from ref.[11]).

An issue that may appear during the process of preparation of the carrier is that the drug added to the mixture may destabilize the previous lyotropic structure. In some cases, to recover the carrier structure, additional components (e.g., a co-surfactant) can be incorporated to the original mixture.

The administration of drugs carried by lyotropics is done via different ways, revealing the broad range of applicability of the method, e.g., intravenous, subcutaneous, oral, transdermal, ophthalmic, and oral [9].

An interesting drug-release mechanism with lyotropics was proposed by Aleandri and co-workers, using light [13]. The lyotropic aqueous mixture was prepared with monoolein and oleic acid, and a guest lipid containing an azobenzene photoactive unit. A limitation of this method is that a transparent medium is needed or, in the case of internal regions, light must be driven (e.g., by optical fibres) to the place where the drug has to be liberated. Figure 11.4 sketches the drug delivery triggered by light. The process is based on the *trans-to-cis* isomerization of the guest lipid, triggered by UV light, and the *cis-to trans* photo-switching induced by blue light. The structural conformation change of the lipid guest allows the permeability of the drug through the main amphiphile double layer. This mechanism is particularly interesting, since it allows control of the dose delivered, being a powerful on-demand treatment method.

A parameter that can be crucial in the stabilization of a particular lyotropic structure is the pH of the mixture. Taking this into account, and also knowing that the pH in different parts of the gastrointestinal tract varies, Negrini and Mezzenga proposed a lyotropic mixture that changes its structure as a function of the pH [14]. The lyotropic mixture was composed of monolinolein/linoleic acid/water, ionic strength 150 mM. In pH 2, characteristic of the stomach conditions, the structure is that of a reverse hexagonal phase and in pH 7, characteristic of the mouth and intestine conditions, the structure is a reverse bicontinuous cubic phase. Since the permeability of a particular drug, across the lyotropic structure, depends on the phase, the mixture can be 'designed' according to the purpose of the treatment. In a proof of concept, it was shown that the diffusion of phloroglucinol (the model drug) and, consequently, the delivery of the active substance, is 4 times faster at pH 7 than at pH 2 [14]. Figure 11.5 depicts the pHs of different regions of the body, and the stabilized structure of the lyotropic nanoparticles.

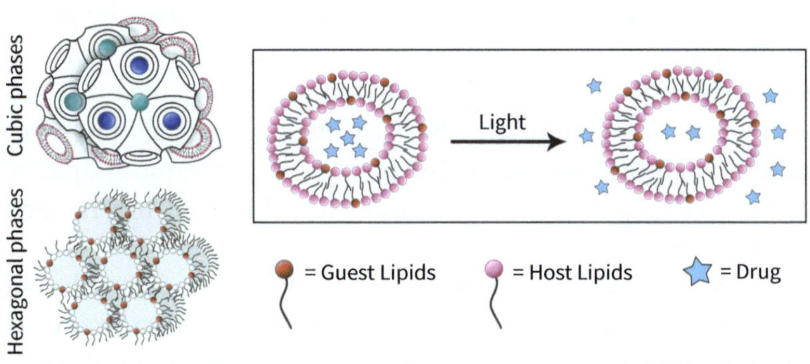

Figure 11.4 Sketch of the light-triggered drug-delivery process.
(Adapted with permission from ref.[13]).

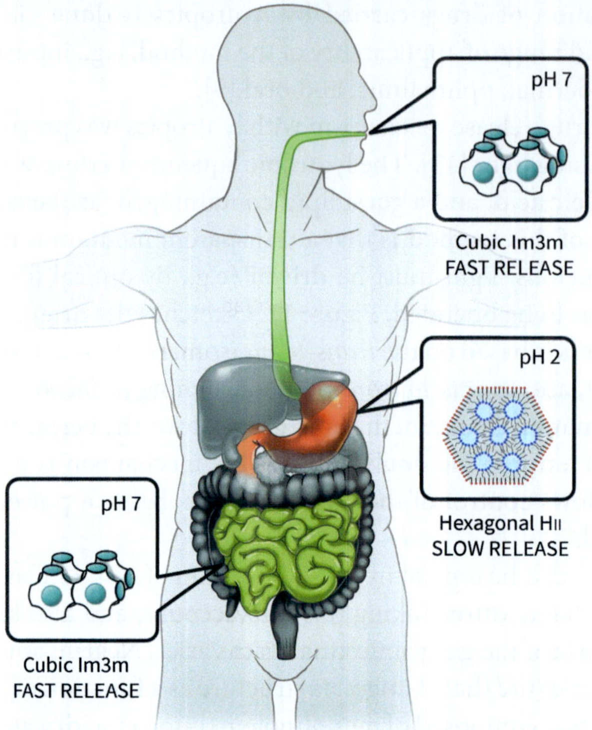

Figure 11.5 Sketch of lyotropic nanoparticle's structure associated with the pH of different parts of the human body.
(Reproduced with permission from ref.[14]).

More recently, lyotropic nanosystems have been employed as vaccine platforms [15]. Von Halling Laier [16] proposed the preparation of hydrated and dried nano-sized cubosomes to deliver vaccines, experimentally demonstrating the method in mice. Rodrigues and co-workers [17] suggested the aqueous mixture of the lipids phytantriol and mannide monooleate, that stabilized hexosomes, as vaccine carriers. The application of cubosomes was also reported as vaccine carriers [18]. During the SARS-COV-2 pandemic, lipid lamellar nanoparticles were used by Moderna and BioNTech/Pfizer to encapsulate vaccines based on mRNA [19].

Lyotropic nanosomes are present in new long-acting sustained parental formulations, not only to deliver nutrients but also pharmaceuticals [20]. Despite the advantages of this application, e.g., pH-dependent delivery, biocompatibility, and delivery control, the subject needs more research to achieve a competitive advantage over other, conventional methods.

Chemotherapy, using an in situ depot of lyotropic mixtures with antitumour drugs, has been employed to treat solid tumours [21]. Drugs are incorporated into the lyotropic mixture which exhibits a lamellar phase. This active fluid is injected directly into the tumour. In that environment, a structural phase transition takes place and the hexagonal mesophase is stabilized. This process allows the controlled

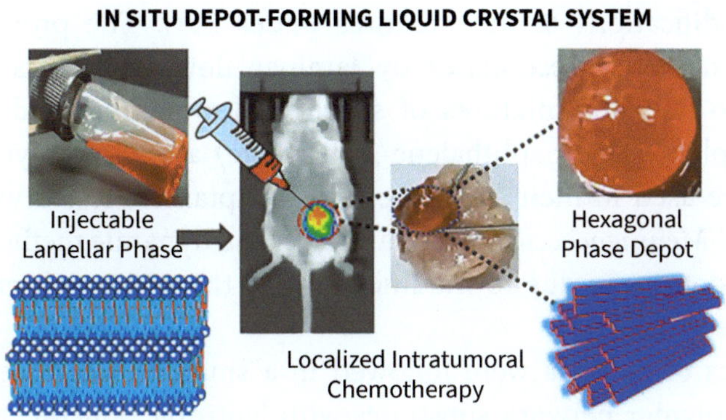

Figure 11.6 Representation of the in situ depot forming methodology of injection directly in a solid tumour. Drugs are incorporated into a lyotropic lamellar liquid crystalline mesophase. After the injection, the lyotropic mixture undergoes a phase transition to the hexagonal mesophase. This process allows the control of the drug delivery time.
(Reproduced with permission from ref.[21]).

delivery of the drug in pre-defined periods varying from a week to one month. Figure 11.6 depicts the in situ depot-forming method.

11.4 Biosensors

Some characteristics of liquid crystals enable them to be employed as sensors of different kinds in the biomedical area [22–24]. Among these characteristics, we may cite the selective reflection of light (e.g., from cholesteric structures), the texture under polarized light (i.e., the topological organization of the symmetry axes of the structure), and the structural phase transitions induced by external stimuli. Another important characteristic of lyotropics is the possibility to prepare biocompatible mixtures, where aqueous mixtures are mainly employed for this purpose.

The application of thermotropic liquid crystalline materials as biosensors was developed in the 1970s, e.g., liquid crystal thermography to detect breast cancer [25]. The mixture employed was cholesteryl oleyl carbonate, cholesteryl nonanoate, and cholesteryl benzoate. The helical pitch of the cholesteric mesophase is particularly sensitive to temperature and temperature gradients, allowing the identification and localization of tumours simply by inspecting the reflected light from regions of the body with higher and lower vascularization.

Lyotropic chromonic aqueous mixtures were shown to be efficient in revealing the presence of bacteria, placed in a liquid crystalline film [26]. The technique uses the fact that the presence of the bacteria distorts the previously aligned director field. These distortions can be followed by observing the textures of the liquid crystalline material, under crossed polarizer optical microscopy.

The detection of macromolecules and microorganisms with lyotropics was proposed by Otón and co-workers [27], using microfluidic devices, exploring the texture modifications of the mixtures under flow. The process of director orientation of nematic mesophases by laminar flow in cells is a well-known procedure. Two aqueous mixtures of sunset yellow FCF (disodium 6-hydroxy 5-[(4-sulfonatophenyl)azo]naphthalene-2-sulfonate) and Cromolyn (sodium cromoglycate) were used in their experiments. The aptamers tested were *Legionella*, *Salmonella*, and *Meningococcus*. An advantage of the dynamic method is the amplification of the detection, due to modifications of the texture observed when the aptamer is present.

The dynamics of bacteria were followed in a smart device that combines the liquid-crystal polymer network substrates with lyotropic aqueous mixture of disodium cromoglycate (DSCG) in the nematic mesophase [28]. The lyotropic mixture is placed in between the substrates, confined to a film of 5 μm thickness. The proof-of-concept was done with the bacterial strain 23857 of *Bacillus subtilis* (length of 5–7 μm and a width of 0.7 μm). Figure 11.7 shows the movement of two bacteria, present in the lyotropic mixture, followed by optical observation of the texture. The uniaxial nematic liquid crystalline director is oriented in the planar geometry, and the bacteria move parallel to the director direction. Due to the birefringence of the liquid crystalline mesophase, the bacteria movement distorts the previously aligned director field and the movement is observed.

The presence of rod-shaped bacteria in biocompatible lyotropic nematic mixtures revealed an interesting coupling between the bacteria and the nematic director [29]. Interestingly, if the bacteria have a flagellum that is employed for propulsion, local melting may occur. This fact can be employed to follow the bacterium's trajectory by simple observation of the liquid crystalline texture under polarized microscopy. Moreover, the spatial director arrangement essentially drives the trajectory of the bacterium. Figure 11.8 shows a texture where the bacteria are aligned along the director direction, evidencing the presence of disclination pairs.

An interesting application of the lyotropic chromonic mixture of disodium cromoglycate (DSCG)/water is to detect cancer biomarker CA125 [30]. The device is composed of two transparent glass slides. The glass surfaces in contact with the lyotropic were coated with a material (polyimide) that forms a thin film. With a convenient rubbing of these films, the cell filled with the liquid crystalline material aligns the molecular chromonic stacks in planar geometry. The method of detection

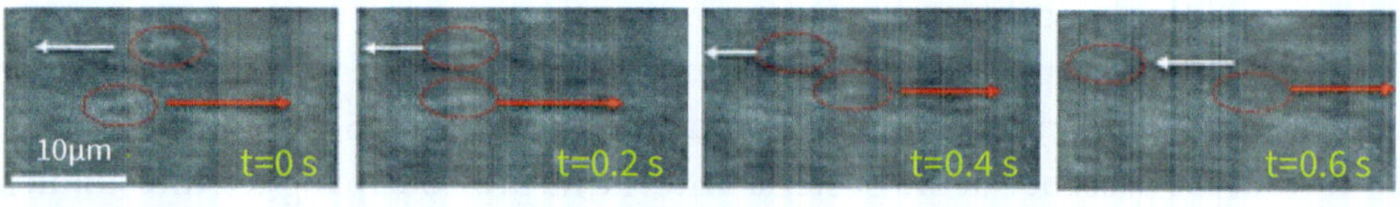

Figure 11.7 Dynamics of two swimming bacteria (circled by red lines) in the lyotropic mixture, in between two liquid-crystal polymer network substrates. White and red arrows indicate the direction of motion and the time is shown in the figure.
(Reproduced with permission from ref.[28]).

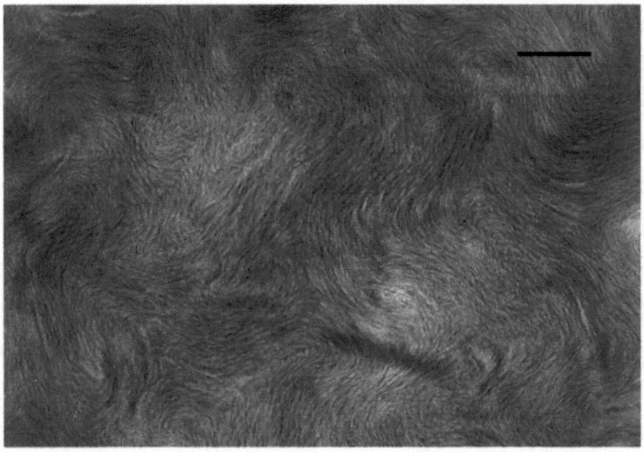

Figure 11.8 *B. subtilis* bacterium in a chromonic biocompatible lyotropic mixture, observed in polarized optical microscopy. The black bar represents 30 μm. Fine white stripes reveal the location of bacteria.
(Reproduced with permission from ref.[29]).

uses the fact that the alignment is distorted by the presence of the cancer biomarker CA125, captured by the anti-CA125 antibody. The cell, filled with the lyotropic mixture, at different (known) concentrations of the biomarker/antibody, was employed in the measurements of the transmitted light intensity. The experimental conditions of the setup allow measurement of samples between crossed or parallel linear polarizers. Calibration curves were established and used to quantify the amount of the biomarkers in a new sample from patients.

Lyotropic liquid crystals, based on the mixture of peptides and a solvent (e.g., water), shows supramolecular arrangements that are promising platforms for biomedical applications [31].

11.5 Synthesis of nanomaterials

This application of lyotropic liquid crystals is based on the complex arrangement of the basic units of the mixture in space. This arrangement, as described in previous chapters, ranges from simple nematics to 3D direct and inverted sponge-like structures. The presence of porous structures allows the use of them as synthetic templates.

An example of this application is the use of tobacco mosaic virus (TMV) to synthesize mesostructured and mesoporous silica, with periodicities of about 20 nm [32]. As previously discussed, these viruses are long cylindrical particles (~300 nm long, ~18 nm wide), having a cylindrical hole of ~4 nm diameter in their centre. The procedure involves the addition of tetraethoxysilane and aminopropyltriethoxysilane in the lyotropic mixture. After the silicification process, the organic part of the sample can be removed.

The space between the basic units (e.g., amphiphilic aggregates) can be used as nano-reactors, to promote chemical reactions in a small and controlled region. An example of this procedure is the synthesis of metals and semiconductors in direct and inverted hexagonal and lamellar structures [33]. The method consists of placing different substances into the lyotropic medium, considering their solubility in the different loci of the structure (hydrophobic and hydrophilic). Different substances are incorporated into different lyotropics, and the reaction is induced by shear mixing of the liquid crystalline materials. The sizes and geometry of the space between the lyotropic aggregates, and the concentrations of the reagents, will define the shape and size of the nanoparticles synthesized. Figure 11.9 shows PbS nanoparticles obtained by shear mixing of lyotropic mixtures (with the corresponding active compounds), in the hexagonal mesophase [33].

Zinc selenide nanostructures were synthesized with a procedure that used an oil-water-amphiphilic block copolymer lyotropic mixture [34]. Spherical domains of lyotropic inverted cubic mesophases were used to grow quantum dots. Cubic and hexagonal direct structures allowed the growth of hollow nanospheres. The presence of spherical and cylindrical aggregates in the lyotropic mixture allowed the growth of hollow nanospheres and hollow nanotubes. Lamellar structures were used to growth free-standing quantum wells.

Lyotropics can be used to construct templates on the nanometre scale and various porous topologies for different purposes [35–37]. Nano- and microparticles of silver were synthesized in the lyotropic-mixture template of triton X-100/water mixture, in the hexagonal and lamellar lyotropic structures [38]. The procedure uses a smaller amount of silver nitrate, and the reducing agent, compared to conventional methods. The shape, and shape anisotropy of particles, could be controlled by the addition of the reducing agent. Moreover, the procedure is performed at room temperature, which is an important advantage compared to traditional methods. Iron-oxide nanoparticles, of polyhedral shape, were synthesized by iron co-precipitation in a lyotropic liquid crystalline medium, in the lamellar and hexagonal

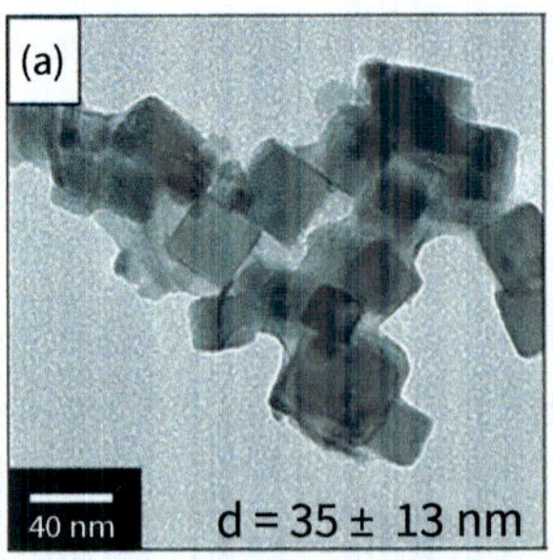

Figure 11.9 Transmission electron microscopy of PbS nanoparticles, synthesized with the method of shear mixing lyotropic mixtures, in the hexagonal mesophase.
(Reproduced with permission from ref.[33]).

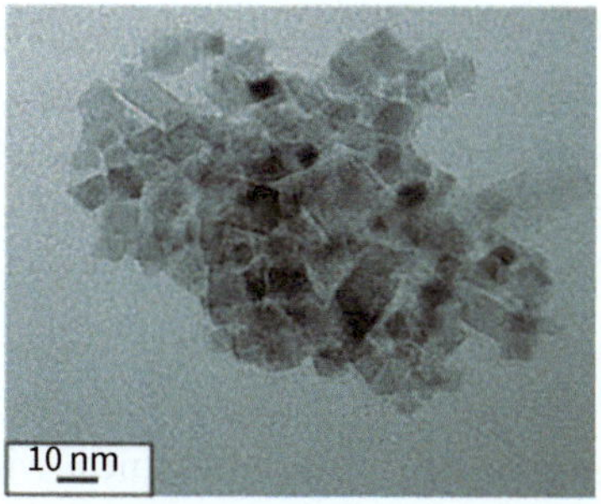

Figure 11.10 Transmission electron microscope images of nanostructures, prepared by co-precipitation of iron (iron chloride precursors, hydrolysed in water with sodium hydroxide) in the lyotropic aqueous mixture of Triton X-45, in the lamellar mesophase.
(Reproduced with permission from ref.[39]).

mesophases [39,40]. Figure 11.10 shows some iron-oxide nanoparticles synthesized via this procedure.

Due to the richness of the spatial arrangements of the molecules in lyotropics, in particular the structures where porosity is present, nanostructures of 0D, 1D, 2D, and 3D may be synthesized *in situ* [41]. Examples of nanostructures synthesized in lyotropic templates are cuprite nanowires, Pt nanotubes, Au microplatelets, 0D nanoparticles, and nanobelts.

11.6 Self-organization in nanotechnology

Chiral plasmonic nanostructures can be fabricated by using cellulose nanocrystals and gold nanorods incorporated into the structure simultaneously [42]. These materials present high circular optical dichroism. The method of preparation of these structures employs the lyotropic mixture of cellulose nanocrystals in water which spontaneously organize in the cholesteric structure, and gold nanorods are introduced into the mixture. After removal of the water, the cholesteric nanostructure with the nanorods incorporated is ready to be used as a helical plasmonic film for further applications. The activity of the nanostructure can be tuned by varying the dimensions (i.e., the shape anisotropy) of the gold nanorods.

Silver nanowires were also incorporated in aqueous mixtures of cellulose nanocrystals to fabricate plasmonic structures, with potential applications in the study of metamaterials [43]. The chiral plasmonic activity may be controlled by modifying the interaction between the silver nanowires themselves and between them and the cellulose nanocrystals.

Lyotropic mixtures were also employed in the production of 'super fibres', or high-performance fibres. These materials present outstanding properties, mainly of tensile strength, fitting to the needs of the aviation industry and personal security (body protection) activities [44]. Fibres (e.g., polybenzoxazole {PBO}, named

Zylon®) are produced by extrusion, where the polymerized material, mixed with the liquid crystalline material, flows through capillaries and is stretched.

Low weight, high mechanical resistance, flexibility, and adequate electrical conductivity are properties desired for fibres employed in the industry for high-technological utilization. The challenge of developing materials with these combined properties is not an easy task, since sometimes a homogeneous mixture of components, each one with a particular needed characteristic, does not occur in the common chemical reactors. At this point, lyotropics, with their nanoscopic structure, may respond to this need. An interesting example of this is the confection of polyaromatic amide (PA) and carbon nanotube (CNT) composite fibres [45]. It was shown that mixtures of PA, CNT, and chlorosulfonic acid exhibit liquid crystalline properties. The homogeneous mixture is then subjected to the extrusion process to produce the fibres. Figure 11.11 shows the texture of the lyotropic mixture, under polarized optical microscopy observation. The birefringence and the Schlieren texture suggest a nematic-like structure. This structure reflects the final properties of the fibres and can be tuned, by changing the relative composition of the mixture, according to the particular application needed.

As a final remark, lyotropics are employed to build devices in the field of soft-matter photonics. The use of thermotropics in optical devices has been known for many years, with remarkable developments for our daily life and in optical research laboratories. On the other side, lyotropics are less employed in the design of optical devices. However, the biocompatible aspect of some lyotropic mixtures accredits them to be used in devices adapted to biological and medical applications, among others. One of the most prominent mixtures in those fields is based on cellulose. Cellulose nanocrystals and cellulose derivatives are natural candidates for biological and medical applications [46]. The helicoidal arrangement of the basic units of the cellulose-based lyotropic mixture is dependent on the relative concentration of the components and the pitch of the helical superstructure can be tuned. Moreover, when dried, the solid film keeps the original helicoidal structure. The employment of this process to build photonic devices uses the fact that the helicoidal structure is

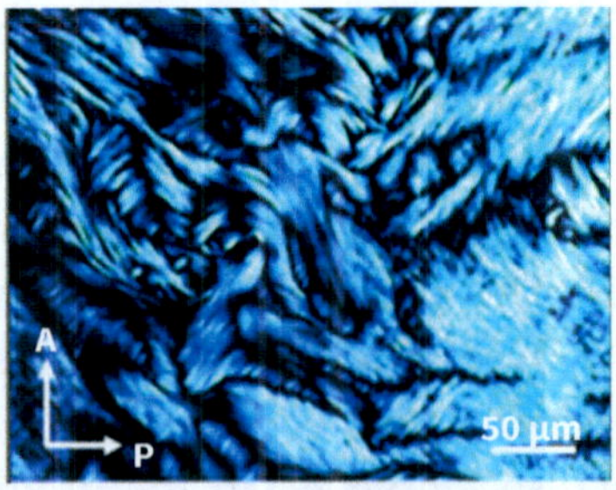

Figure 11.11 Texture of the lyotropic mixture of polyaromatic amide, carbon nanotubes, and chlorosulfonic acid, observed in the polarized optical microscope. P and A represent the orientation of the polarizer and analyser.
(Reproduced with permission from ref.[45]).

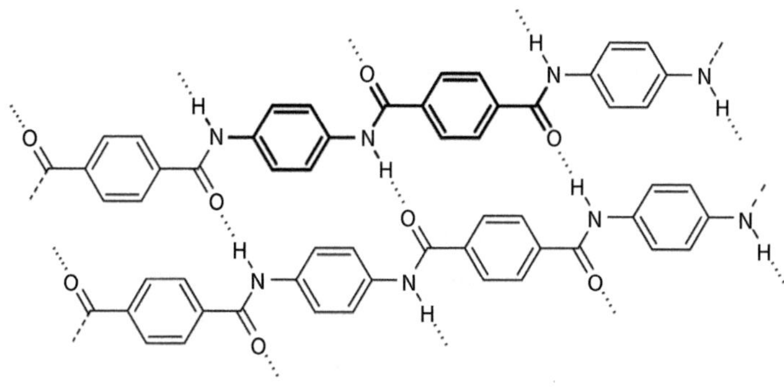

Figure 11.12 Arrangement in the molecular plane of a Kevlar fibre.

sensitive to humidity, solvents, gases, external strain, and temperature. Tuning the helicoidal structure to pitches in the visible-light range allows simple observation of the wavelength of the reflected light by the structure.

Lyotropic liquid crystals made of polymers dissolved in a solvent can be a precursor of high technology materials, e.g., Kevlar. This is an aramid synthetic fibre produced from a mixture of polymers (Figure 11.12) and a solvent, subjected to a drying process [47]. The process is similar to that used by spiders to engender a silk web. The resistance of Kevlar against mechanical stress and impact is very high, allowing its use in bullet-proof protective clothing and fire-resistant materials.

References

[1] R. Mazzenga, J. M. Seddon, C. J. Drummond, B. J. Boyd, G. E. Schröder-Turk, and L. Sagalowicz (2019). Nature-inspired design and application of lipidic lyotropic liquid crystals. *Adv. Mater.*, 31, 1900818.

[2] I. Martiel, N. Baumann, J. J. Vallooran, J. Bergfreund, L. Sagalowicz, and R. Mezzenga (2015). Oil and drug control the release rate from lyotropic liquid crystals. *J. Controlled Release*, 204, 78–84.

[3] L. C. Maillard (1912). Formation of melanoidins in a methodical way. *Comptes Rendus*, 154, 66–68.

[4] A. Sadeghpour and M. Rappolt (2016). Lyotropic liquid crystalline phases for the formulation of future functional foods. *J. Nutr. Health Food Eng.*, 5, 553–557.

[5] H. Iwai, J. Fuksawa, and T. Suzuki (1998). A liquid crystal application in skin care cosmetics. *Int. J. Cosmetic Science*, 20, 87–102.

[6] P. Hong-in and W. Chaiyana (2022). Potential cosmeceutical lamellar liquid crystals containing black longan (*Dimocarpus longan* Lour.) seed extract for MMP-1 and hyaluronidase inhibition. *Sci. Rep.*, 12, 7683.

[7] C. Yhirayha, S. Soontaranon, S. Wittaya-Areekul, and T. Pitaksuteepong (2014). Formulation of lyotropic liquid crystal containing mulberry stem extract: influences of formulation ingredients on the formation and the nanostructure. *Int. J. Cosmetic Science*, 36, 213–220.

[8] C. Guo, J. Wang, F. Cao, R. J. Lee, and G Zhai (2010). Lyotropic liquid crystal systems in drug delivery. *Drug Discovery Today*, 15, 1032–1040.

[9] D. H. Kim, A. Jahn, S. J. Cho, J. S. Kim, M. H. Ki, and D. D. Kim (2015). Lyotropic liquid crystal systems in drug delivery: a review. *J. Pharmaceutical Investigation*, 45, 1–11.

[10] D. Demurtas, P. Guichard, I. Martiel, R. Mezzenga, C. Hébert, and L. Sagalowicz (2015). Direct visualization of dispersed lipid bicontinuous cubic phases by cryo-electron tomography. *Nature Communications*, 6, 8915.

[11] J. Kuntsche, J. C. Horst, and H. Bunjes (2011). Cryogenic transmission electron microscopy (cryo-TEM) for studying the morphology of colloidal drug delivery systems. *International Journal of Pharmaceutics*, 417, 120–137.

[12] M. L. Lynch, A. Ofori-Boateng, A. Hippe, K. Kochvar, and P. T. Spicer (2003). Enhanced loading of water-soluble actives into bicontinuous cubic phase liquid crystals using cationic surfactants. *J. Colloid Interf. Sci.*, 260, 404–413.

[13] S. Aleandri, C. Speziale, R. Mezzenga, and E. M. Landau (2015). Design of light-triggered lyotropic liquid crystal mesophases and their application as molecular switches in 'on demand' release. *Langmuir*, 31, 6981–6987.

[14] R. Negrini and R. Mezzenga (2011). pH-responsive lyotropic liquid crystals for controlled drug delivery. *Langmuir*, 27, 5296–5303.

[15] M. Chountoulesi, S. Pispas, I. K. Tseti, and C. Demetzos (2022). Lyotropic liquid crystalline nanostructures as drug delivery systems and vaccine platforms. *Pharmaceuticals*, 15, 429.

[16] C. von Halling Laier, B. Gibson, M. van de Weert, B. J. Boyd, T. Rades, A. Boisen, S. Hook, and L. H. Nielsen (2018). Spray dried cubosomes with ovalbumin and Quil-A as a nanoparticulate dry powder vaccine formulation. *International Journal of Pharmaceutics*, 550, 35–44.

[17] L. Rodrigues, K. Kyriakos, F. Schneider, H. Dietz, G. Winter, C. M. Papadakis, and M. Hubert (2016). Characterization of lipid-based hexosomes as versatile vaccine carriers. *Mol. Pharmaceutics*, 13, 3945–3954.

[18] Z. Liu, L. Luo, S. Zheng, Y. Niu, R. Bo, Y. Huang, J. Xing, Z. Li, and D. Wang (2016). Cubosome nanoparticles potentiate immune properties of immunostimulants. *Int. J. Nanomedicine*, 11, 3571–3583.

[19] R. Tenchov, R. Bird, A. E. Curtze, and Q. Zhou (2021). Lipid nanoparticles—from liposomes to mRNA vaccine delivery, a landscape of research diversity and advancement. *ACS Nano*, 15, 16982–17015.

[20] V. P. Chavda, S. Dawre, A. Pandya, L. K. Vora, D. H. Modh, V. Shah, D. J. Dave, and V. Patravale (2022). Lyotropic liquid crystals for parenteral drug delivery. *Journal of Controlled Release*, 349, 533–549.

[21] R. Saklani, P. K. Yadav, M. A. Nengroo, S. L. Gawali, P. A. Hassan, D. Datta, D. P. Mishra, I. Dierking, and M. K. Chourasia (2022). An injectable in situ depot-forming lipidic lyotropic liquid crystal system for localized intratumoral drug delivery. *Mol. Pharmaceutics*, 19, 831–842.

[22] V. K. Gupta, J. J. Skaife, T. B. Dubrovsky, and N. L. Abbott (1998). Optical amplification of ligand-receptor binding using liquid crystals. *Science*, 279, 2077–2080.

[23] S. Norouzi and M. Sadati (2020). Liquid crystal-based biosensors annals of biostatistics & biometric applications. *Annals of Biostatistics & Biometric Applications*, 4. doi: 10.33552/ABBA.2020.04.000590.

[24] C. Luan, H. Luan, and D. Luo (2020). Application and technique of liquid crystal-based biosensors. *Micromachines*, 11, 176.

[25] T. W. Davison, K. L. Ewing, J. Fergason, M. Chapman, A. Can, and C. C. Voorhis (1972). Detection of breast cancer by liquid crystal thermography. A preliminary report. *Cancer*, 29, 1123–1132.

[26] S. V. Shiyanovskii, O. D. Lavrentovich, T. Schneider, T. Ishikawa, I. I. Smalyukh, C. J. Woolverton, G. D. Niehaus, and K. J. Doane (2005). Lyotropic chromonic liquid crystals for biological sensing applications.*Mol. Cryst. Liq. Cryst.*, 434, 259–270 [587–598].

[27] E. Otón, J. M. Otón, M. Caño-García, J. M. Escolano, X. Quintana, and M. A. Geday (2019). Rapid detection of pathogens using lyotropic liquid crystals. *Optics Express*, 27,10098–10107.

[28] N. P. Dhakal, J. Jiang, Y. Guo, and C. Peng (2020). Self-assembly of aqueous soft matter patterned by liquid-crystal polymer networks for controlling the dynamics of bacteria. *ACS Appl. Mater. Interfaces*, 12, 13680–13685.

[29] S. Zhou, A. Sokolov, O. D. Lavrentovich, and I. S. Aranson (2014). Living liquid crystals. *PNAS*, 111, 1265–1270.

[30] H. Shaban, M.-J. Lee, and W. Lee (2021). Label-free detection and spectrometrically quantitative analysis of the cancer biomarker CA125 based on lyotropic chromonic liquid crystal. *Biosensors*, 11, 271.

[31] A. Boruaha and A. Roy (2022). Advances in hybrid peptide-based self-assembly systems and their applications. *Biomater. Sci.*, 10, 4694–4723.

[32] C. E. Fowler, W. Shenton, G. Stubbs, and S. Mann (2001). Tobacco mosaic virus liquid crystals as templates for the interior design of silica mesophases and nanoparticles. *Advanced Materials*, 13, 1266–1269.

[33] T. M. Dellinger and P. V. Braun (2004). Lyotropic liquid crystals as nanoreactors for nanoparticle synthesis. *Chem. Mater.*, 16, 2201–2207.

[34] G. N. Karanikolos, P. Alexandridis, R. Mallory, A. Petrou, and T. J. Mountziaris (2006). Water-based synthesis of ZnSe nanostructures using amphiphilic block copolymer stabilized lyotropic liquid crystals as templates. *Nanotechnology*, 17, 3121–3128.

[35] G. S. Attard, C. G. Goltner, J. M. Corker, S. Henke, and R. H. Templer (1997). Liquid-crystal templates for nanostructured metals. *Angew. Chem. Int. Ed. Engl.*, 36, 1315–1317.

[36] M. Antonietti and G. A. Ozin (2004). Promises and problems of mesoscale materials chemistry or why meso? *Chem. Eur. J.*, 10, 28–41.

[37] C. Wang, D. Chen, and X. Jiao (2009). Lyotropic liquid crystal directed synthesis of nanostructured materials. *Sci. Technol. Adv. Mater.*, 10, 023001.

[38] S. Umadevi, R. Umamaheswari, and V. Ganesh (2017). Lyotropic liquid crystal-assisted synthesis of micro-and nanoparticles of silver. *Liquid Crystals*, 44, 1409–1420.

[39] M. Worden, M. A. Bruckman, M.-H. Kim, N. F. Steinmetz, J. M. Kikkawa, C. LaSpina, and T. Hegmann (2015). Aqueous synthesis of polyhedral 'brick-like' iron oxide nanoparticles for hyperthermia and T 2 MRI contrast enhancement. *J. Mater. Chem. B Mater. Biol. Med.*, 3, 6877–6884.

[40] S. M. Salili, M.Worden, A. Nemati, D. W. Miller, and T. Hegmann (2017). Synthesis of distinct iron oxide nanomaterial shapes using lyotropic liquid crystal solvents. *Nanomaterials*, 7, 211.

[41] C. Wang, D. Chen, and X. Jiao (2009). Lyotropic liquid crystal directed synthesis of nanostructured materials. *Sci. Technol. Adv. Mater.*, 10, 023001.

[42] A. Querejeta-Fernańdez, G. Chauve, M. Methot, J. Bouchard, and E. Kumacheva (2014). Chiral plasmonic films formed by gold nanorods and cellulose nanocrystals. *J. Am. Chem. Soc.*, 136, 4788–4793.

[43] G. Chu, X. Wang, T. Chen, J. Gao, F. Gai, Yu Wang, and Y. Xu (2015). Optically tunable chiral plasmonic guest–host cellulose films weaved with long-range ordered silver nanowires. *ACS Appl. Mater. Interfaces*, 7, 11863–11870.

[44] Y. Teramoto and F. Kubota (2016). Zylon®: Super Fiber from Lyotropic Liquid Crystal of the Most Rigid Polymer. In: Japan The Society of Fiber Science and Techno (eds), *High-Performance and Specialty Fibers, Concepts, Technology and Modern Applications of Man-Made Fibers for the Future*, chapter 11, pp 191–216.

[45] K. H. Ryu, J. G. Kim, D. Lee, S. G. Kim, B. C. Ku, J. Y. Hwang, K. U. Jeong, N. D. Kim, and D. Y. Kim (2023). Boost up the mechanical and electrical property of CNT fibers by governing lyotropic liquid crystalline mesophases with aramid polymers for robust lightweight wiring applications. *Advanced Fiber Materials*, 5, 514–526.

[46] L. L. Ma, C. Y. Li, J. T. Pan, Y. E. Ji, C. Jiang, R. Zheng, Z. Y. Wang, Y. Wang, B. X. Li, and Y. Q. Lu (2022). Self-assembled liquid crystal architectures for soft matter photonics. *Light: Science & Applications*, 11, 270.

[47] R.-M. Wang, S.-R. Zheng, and Y.-P. Zheng (2011). *Polymer Matrix Composites and Technology: A Volume in Woodhead Publishing Series in Composites Science and Engineering*. Cambridge: Woodhead Publishing.

Index